Photoshop

CC ESSENTIALS FOR PHOTOGRAPHERS:

Chelsea & Tony Northrup's video book

Published by:

Mason Press, Inc.
139 Oswegatchie Rd.
Waterford, CT 06385

ISBN: 978-0-9979505-0-2

Printed and bound in the United States of America by Signature Book Printing, *www.sbpbooks.com*

Editors: Tanya Egan Gibson, Justin Eckert, Chelsea Northrup
Designer: Chelsea Northrup
Producers: Justin Eckert, Siobhan Midgett

Trademarks

For Eric & Stephanie Lowenbach

Your friendship & coffee brewing skills made many of these pictures possible.

mojavemorning.com

TABLE OF CONTENTS

20 Configuring Preferences

ACKNOWLEDGEMENTS

First, thank you to every reader! We owe an extra thanks to everyone who ordered this book during the pre-order phase and patiently waited while we got it perfect for publication. Our dream is to spread the art of photography as far as possible throughout the world, and it we couldn't do it without you.

Creating this has been a massive effort requiring the skills of an expert team. In no particular order:

- Justin Eckert, video producer and overall bro
- Siobhan Midgett, video producer and close friend
- Tanya Gibson, editor and proofreader
- Phil Nanzetta and Linda Wood (at Signature Book Printing), our printer

Please Read This!

Please read this! I'll tell you how to get free stuff and how to use this book.

First, visit *sdp.io/psedl* to watch the videos and download the sample images.

Optionally, to receive the free updates, access the Facebook group, and the free ebook, register your purchase at *sdp.io/register*. You don't have to register to see the videos or download the sample files. If you have a problem with anything, send me a message at *tony@northrup.org*.

This book includes many unique benefits:

- **Video training**. This book includes more than 10 hours of Photoshop video training. Visit *sdp.io/psedl* to view the videos. You can watch the videos instead of reading the book, but you'll learn best if you do both. We find videos are more effective for teaching artistic techniques. The book format is more effective for teaching technical processes and for information you might want to scan or reference, such as preferences.

- **Mentoring**. Optionally, join our private Facebook group by registering at *sdp.io/register*. Post your questions or photos for help and feedback from Tony, Chelsea, and other readers.

- **Lifetime updates.** I plan to update this book every time Adobe adds major new features to Photoshop CC. Unlike Adobe, I won't charge a monthly fee; I'll give those updates to you for free.

- **Sample images.** Visit *sdp.io/psedl* to download every picture you see in this book. You'll learn Photoshop faster if you walk through every step-by-step process in this book using either our sample images or your own images. Because we're photographers, Chelsea & I took every picture in the book; we don't use stock photography.

- **Free ebook.** We'll send you a link to download your ebook when we process your registration. Keep it on your tablet or smartphone for quick reference when you don't have the book handy.

Chelsea and I have a weekly live show on YouTube about photography that includes our reviews of actual reader photos. To watch it, and new tutorial videos when we release them, visit *sdp.io/yt* and subscribe to our YouTube channel.

If you want to see new photos that we make, check out our social media:

- **YouTube**: *youtube.com/c/TonyNorthrup*

- **Facebook**: *fb.com/NorthrupPhotography*

- **Twitter**: *@TonyNorthrup and @ChelseaNorthru*

- **Instagram**: *@TonyNorthrup and Chelsea_Northrup*

- **Snapchat**: *@theTonyNorthrup and @ChelseaNorthrup*

We can't answer every Photoshop question, so you're more likely to get a response if you post it in the private Facebook group. If you have a problem with the book, email me at *tony@northrup.org*.

One request: when someone compliments your pictures and editing, tell them you learned from *Chelsea & Tony Northrup's Photoshop Essentials*. We've spent more than a year creating this book, and we hope to spend the rest of our lives improving it and supporting readers. Making great images and helping people is what we love to do. If we're going to make this work, however, we need your help spreading the word.

1 QUICK START

Watch training videos at:
SDP.io/PSEDL

This chapter will get you started in Photoshop as quickly as possible. Most of these topics are covered in far more detail later in this book.

As with most parts of this book, you can choose to either read the chapter or watch the associated video training, depending on which method is most effective for your personal learning style. Either way, you'll get the best results when you follow along in Photoshop on your own computer, using either our sample files or your own pictures.

Using This Book

You'll often need to select menu commands. The menu system is at the top of the Photoshop window—it shows **File**, **Edit**, **Image**, **Layer**, **Type**, **Select**, **Filter**, **3D**, **View**, **Window**, and **Help** (though those items can vary). I'll show you menu choices by separating the items with vertical bars, such as **File | New** or **Image | Image Rotation | Flip Canvas Horizontal**.

The tools are located on the left side of your window by default. You might see one long row, or two shorter rows (shown next), depending on your screen size. If you see only one row, the tools might be in a different order. Just look for the matching icon. Click-and-hold an icon to open that tool's sub-menu, also shown next.

My examples were created with a Windows PC, but you won't have any problem using a Mac. The application is essentially identical on both PC and Mac. However, depending on the keyboard, Mac users might see the **Alt** key labeled as **Opt** or the **Ctrl** key labeled as **Cmd**.

There are dozens of different ways to perform any task in Photoshop. I'll show you one or two ways, but you might stumble across a way that works better for you. There's not any right or wrong way; just do what gets the job done!

Buying Options

If you don't yet have Photoshop, Adobe offers a free 30-day trial at *photoshop.com*. After that, you'll have to pay. You have two choices:

- **Lease Photoshop CC**. For $10 per month in the US, or substantially more in some other countries, you can lease Photoshop CC and Lightroom CC at *sdp.io/adobedeal* or directly from Adobe. It might seem strange to lease software, but it's the best deal available, and you'll always have free updates.

- **Buy Photoshop CS6**. Technically, Adobe doesn't sell Photoshop anymore. However, you can find copies of the last version of Photoshop, CS6, on eBay or other outlets that sell used products.

Besides free updates, Creative Cloud (CC) also gives you access to some online features, like file sharing. While we don't find those features useful at the moment, we have Creative Cloud licenses for everyone in the office, which is what I recommend to everyone.

Computer Requirements

Almost any modern computer can run Photoshop, including all recent versions of Windows and Mac OS. You don't need a particularly powerful computer for most photography tasks. However, if you get a high-megapixel camera (such as a 36-megapixel or 50-megapixel camera), Photoshop will definitely become sluggish on slower computers.

If you create images with multiple layers, or combine multiple images together (such as when doing image stacking or creating a panorama), Photoshop will be slow even on powerful computers. I regularly stack a dozen 50-megapixel raw files and create multi-gigapixel panoramas. There's no computer I can buy that will make that fast.

The point is, almost any computer will run Photoshop, and performance will be fine for all simple tasks. If you get into big images and complex layering, you'll simply have to be patient, no matter what your computer budget is.

If Photoshop seems slow and you want it to be faster, these tips can help:

- **Add more memory (RAM).** Photoshop runs fine with 4 GB of RAM, but 8 GB definitely helps. I have 64 GB of RAM in my computer, and while most people won't need that, Photoshop definitely makes use of it with my gigapixel panoramas.

- **Use a Solid State Drive (SSD)**. Photoshop reads your images files from your hard drive. It also uses the hard drive when it runs out of RAM. Therefore, a fast hard drive can make Photoshop faster. SSDs are much faster than conventional, magnetic drives, and the easiest way to make Photoshop faster is to buy a computer with an SSD. If you're skilled with computers and you have room for an extra drive, you can also add a new SSD to your computer and then configure that SSD drive as your scratch drive (**Edit | Preferences | Scratch Disks** on Windows or **Photoshop CC | Preferences | Scratch Disks on a Mac**).

Using Lightroom and Photoshop Together

Adobe Lightroom is designed to work closely with Photoshop. Use Lightroom to import, organize, and export your photos, as well as for light editing (such as cropping). For more serious editing, such as retouching portraits and working with layers, open the files in Photoshop.

To open an image in Photoshop from Lightroom, right-click the image in the **Grid** view, select **Edit In**, and then select **Edit in Adobe Photoshop**, as shown next.

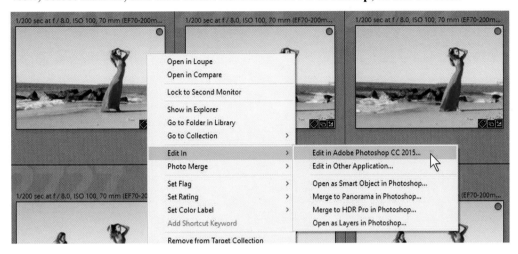

When you're done with Photoshop, simply close the file, and the edits should automatically appear as a new image in Lightroom. If you don't see your picture in Lightroom, be sure to select **All Photographs** under **Catalog**. Edited photos won't appear if you have Previous Import selected because they weren't part of the previous import.

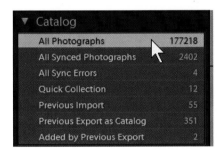

Tip: To use Lightroom to quickly find photos you've edited in Photoshop, use the Text filter to search for .TIF. That's the most common file extension for edited files.

Notice that you can also open multiple pictures in Photoshop from Lightroom by selecting **Merge to Panorama**, **Merge to HDR Pro**, or **Open as Layers**. The usefulness of those options will become clear as you work through this book.

Cropping

As a rule, I always rotate and crop my images in Lightroom, because Lightroom makes it easy to crop wider if I need to (for example, when making a print that requires space on the edges that will be covered by a frame). However, Photoshop CC 2015.5 introduced a cropping feature that Lightroom doesn't yet have: Content-aware Crop.

If you rotate and level a picture with a traditional cropping tool, the image must get smaller because the cropping tool would cut off the corners. With Content Aware crop, demonstrated by the next two figures, Photoshop tries to fill in the corners of the image so you don't have to change the resolution.

Of course, using Content Aware crop is asking Photoshop to fill in parts of your picture that your camera didn't capture. Therefore, it's often not perfect. However, it often works perfect, and can save an image that would otherwise be cropped too tightly after levelling.

Zooming & Panning

You'll be constantly zooming in to see the details of your photos while you edit in Photoshop. The quickest way to zoom in and out is to hold down the **Alt** or **Opt** key and then use the scroll wheel on your mouse. If you don't want to hold down the **Alt** or **Opt** key, choose **Edit | Preferences | Tools** (on Windows) or **Photoshop CC | Preferences | Tools (on a Mac)** and select the **Zoom with Scroll Wheel** checkbox.

Once zoomed, you won't be able to see your entire picture. To quickly pan around the picture, hold down the Space bar. Photoshop changes your cursor to a hand, which you can use to drag the picture around the screen.

Resetting Settings

If you mess something up and want to return to the default settings in a dialog, you can usually hold down the Alt key (on a PC) or the Opt key (on a Mac). This often changes the **Cancel** button to **Reset**, as shown next. Click **Reset** to return to the default settings.

Workspaces

Photoshop has far too many tools, and it's easy to be overwhelmed. Workspaces show and hide tools, so you see only exactly what you need for your current task. They can also be customized to show those tools you personally use most often.

By default, Photoshop has the **Essentials** workspace selected. However, I prefer using the **Photography** workspace. To choose the Photography workspace, select the **Window** menu, select **Workspace**, and then select **Photography** (shown next).

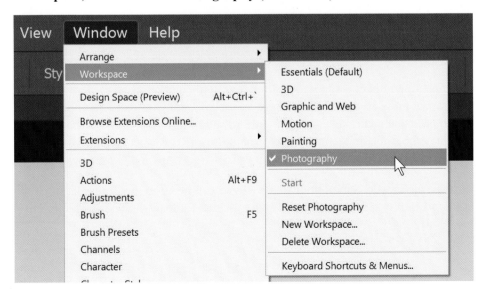

Photoshop will remember any changes you make to the layout of your workspace after you close the app. If you want to permanently remember your current workspace, select **Window | Workspace | New Workspace**. If you want to go back to Photoshop's original settings, select **Window | Workspace| Reset Photography**.

Canvas Size & Image Size

Photographers often need to resize images. Sometimes, a website will require an image to be exactly 500 pixels or 1000 pixels wide. Other times, you might need to crop an image to exactly 8x10.

These tasks are much more natural in Lightroom using the Crop and Export tools. However, if you must do them in Photoshop, it can be done.

To save an image at a specific resolution, select **File | Export | Export As**. On the Export As dialog (shown next) set the **Width** or **Height** to the required resolution. Changing one will automatically change the other; you shouldn't use this tool to change the aspect ratio of the picture.

The aspect ratio is the shape of the image—the width compared to the height. Most cameras take pictures with a 2x3 aspect ratio, so that an image printed at 2x3", 4x6", or 8x12" would be uncropped. If you were to print such an image at 8x10", you would need to crop two inches off the width. If you were to display the same image in a video (which uses a 16:9 aspect ratio), you would need to crop 15% from the height because a 16:9 TV has a 9x16 aspect ratio and is much wider.

Use the Crop tool to change the aspect ratio of your image. To select the Crop tool, press C, or click the Crop tool on the Tools panel. By default, it's the fifth item on the toolbar, shown selected in the following example.

The toolbar directly beneath the menu bar changes depending on the tool you have selected. When you select the Crop tool, you can use the toolbar to change the aspect ratio of the image and select which parts of the image are cropped, as shown next.

Adjustment Layers

The best way to make overall adjustments to your photo, such as contrast, brightness, and saturation, is to use the Develop module in Lightroom. Lightroom is fast, easy, and always reversible.

You can make similar adjustments in Photoshop using adjustment layers. First, make sure you're using the Photography workspace. In the Adjustments panel, select the type of adjustment you want to make, as shown next.

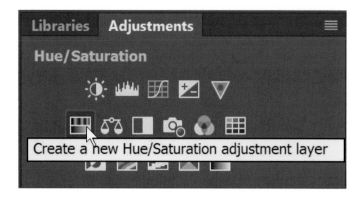

This creates a new layer on top of your existing layer (shown in the lower-right corner of the following example) and opens the Properties panel so that you can adjust the settings. The example shows the Hue/Saturation adjustment with **Reds** selected to change the color of the dress (and the model's lips) to green.

There are a wide variety of adjustment layers available. Spend some time playing with them. Later in this book, we'll dig into detail about how to use each adjustment. For now, know that you can click the reset button at the bottom of the panel (shown next) to erase your changes.

Using Brushes

Even though you're using Photoshop for photography, rather than painting, you'll find brushes incredibly useful. Photoshop uses brushes for everything from removing objects to adding snow.

You won't often paint directly on a photograph, but it's the easiest way to familiarize yourself with how brushes work. Here's a step-by-step example:

1. If you created an adjustment layer, delete it, or select your background layer.
2. Select the brush tool (shown next).
3. Now, select your foreground color by clicking the top box at the bottom of the Tools panel, shown next.

4. Next, use the Color Picker to select a color you want to draw with. You can click anywhere in your photo to sample a color.

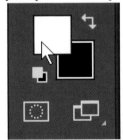

5. Now, draw on your photo. Use the **[** and **]** keys to make your brush bigger or smaller.

You can probably do better than that. Try the Brush toolbar at the top of the screen—it's another way to change the size of the brush, but more importantly, you can adjust the brush hardness. A hard brush has sharp edges that will rarely look natural in a photograph. A soft brush fades the edges into the surrounding image. As you edit photos, you'll frequently be adjusting the hardness of your brushes up or down to create different effects.

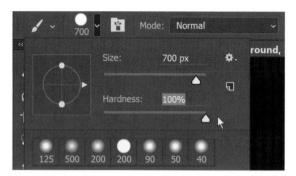

History and Undo

Maybe you don't like the drawing you just made on top of your picture. Don't worry, you can always undo it by repeatedly pressing **Alt+Ctrl+Z** (on a Windows PC) or **Opt+Ctrl+Z** (on a Mac).

You can also view a list of all the changes you've made by selecting **Window | History**. Scroll up, and select the last change you made before the first **Brush Tool** change.

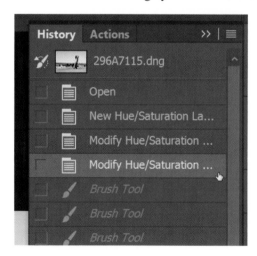

To return to your original photo, click the top item in the list.

Healing

Here's the most common brush use for photographers: healing. Select the Spot Healing Brush Tool by clicking and holding the healing tool on the toolbar until the full list appears. Most tools provide multiple options if you hold them down.

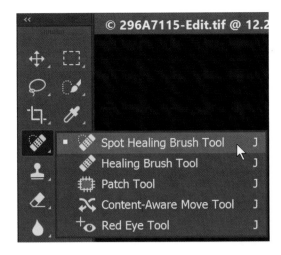

Now, use the [and] keys to make your healing brush large enough to remove an object. Here, I painted the healing brush over one of my yellow brush strokes, and Photoshop immediately removed it.

If Photoshop doesn't completely remove the object, or it puts something strange in your picture, just repeat the process and remove the remaining artifacts.

The healing brush isn't perfect, but it's pretty good.

Cloning

Before the healing brush existed, we photographers had to rely on cloning to copy another part of the picture over an object we wanted to remove. Today, most object and flaw removal is best done with the healing brushes. However, cloning still has some uses.

For example, imagine that I want to add another copy of Emily to this photo. First, I'll select the Clone Stamp Tool, shown next.

Then, I'll make a big, soft brush. The soft edges will assure that the clone blends more naturally with the background.

Now, hold down the **Alt** or **Opt** key and click your clone source—this is what you want to make a second copy of. Once you've established your source, paint in the second copy of it, as shown next.

Dodging and Burning

Dodging makes parts of your picture brighter, while burning makes parts darker. These terms make perfect sense if you have darkroom experience making prints with an enlarger from a negative. If not, they make no sense at all. The term "burning" would seem to make the picture brighter, but it does just the opposite.

As shown next, the Dodge tool looks like a Popsicle, while the Burn tool looks like someone making a circular shape with their hand. Seriously, some younger learners might think the hardest part of learning Photoshop is learning its outdated terminology and iconology, which are rooted in 30-year-old practices.

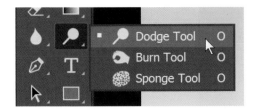

Once you select a tool, select **Shadows**, **Midtones**, or **Highlights** for **Range** on the toolbar (shown next). **Shadows** applies the tool to the darkest parts of your picture, while **Highlights** applies it to the brightest parts.

In the next example, I burned the shadows in Emily's clone (on the left) to make them darker. I dodged the highlights to make them brighter. By making the shadows darker and the highlights brighter, I increased the contrast in her dress.

Transforms

You can use the Transform tool to change the shape of an image or layer. First, select your entire image by pressing **Ctrl+A**. Then, select **Edit | Free Transform**. The Free Transform tool is very powerful. Try holding down the Ctrl key and dragging each of the corners. That's an easy way to straighten pictures of buildings taken with a wide angle lens, as this before-and-after example shows.

If a picture isn't level, hover your cursor outside the edges of the picture until your cursor becomes a double arrow. Then, drag your mouse to rotate the picture.

Take some time to experiment with the effects of dragging the sides and corners while alternately holding down the **Ctrl**, **Shift**, and **Alt** keys:

- **Ctrl (or Cmd)**. Transform the picture into something non-rectangular.
- **Shift**. Lock the aspect ratio so the picture doesn't become tall or wide.
- **Alt**. Lock the picture at its center point instead of the opposite corner.

You can hold down more than one of those keys at a time to combine the effects.

Selection Tools

Selecting parts of a picture in Photoshop allows you to copy and remove objects, as well as apply effects to only specific parts of a picture. Photoshop uses a moving dotted line to show you your current selection; this is commonly known as the "marching ants."

To select the entire picture or layer, press **Ctrl+A** or **Select | All** from the menu. You'll use this often when you need to move or change the entire image, so it's worth memorizing the keyboard shortcut.

To deselect everything, press **Ctrl+D** or **Select | Deselect** from the menu. That's another task that's common enough to warrant memorizing the keyboard shortcut. You'll often need to clear your current selection in order to create a new selection.

To select parts of the picture, use the Quick Selection Tool, shown next.

Once you've selected the tool, click-and-drag over the parts of the picture that you want to select. In the following example, I dragged the mouse near the edges of the mountain, and Photoshop selected those edges automatically. Photoshop missed a few spots, so I clicked-and-dragged again to fill in those spots. Selecting part of the picture is useful for making parts of your picture brighter or darker, fixing selective color problems, or removing objects.

If you accidentally select too much, don't worry—you can always undo it. Hold down the **Alt** or **Opt** key, and then click the parts of the picture that you want to deselect. In the following example, I held down the **Alt** key and dragged inside the shadowed part of the mountain to remove that part of the selection. This would be useful if I wanted my adjustments to not change that part of the mountain.

Filters

Photoshop includes many practical and fun filters that apply special effects to your image. You can use the Filter Gallery (**Filter | Filter Gallery**) to browse more artistic filters.

If the Filter | Filter Gallery menu option isn't available for your current picture, follow these steps to convert it ot the right format:

1. Select **Image | Mode | 8 Bits/Channel**.
2. Select **Image | Mode | RGB Color.**
3. Select **Filter | Convert for Smart Filters** and click **OK**.
4. Now, you should be able to select **Filter | Filter Gallery**.

As shown next, you can browse and adjust dozens of filters to create different effects.

When you open the Filter Gallery, your image might be zoomed in to 100%. You can zoom out using the controls in the lower-left corner, as shown next. You can also hold down the **Alt** or **Opt** key and zoom using the scroll wheel.

Remember, if you want to reset the adjustments, hold down the Alt or Opt key to change the Cancel button to Reset. Then, click Reset.

Liquify

Ready for fun? Open a picture of a friend and select **Filter | Liquify**. We'll cover how to get great results from it later, but for now, just have fun giving your friend big eyes and a small body, or vice-versa. Try the different tools along the left side of the picture.

With practice, you'll be able to use Liquify to fix wrinkles and folds in outfits, or change someone's face and body however you need to. Its capabilities are both powerful and controversial.

Layers

Layers apply effects to small parts of a picture and allow you to selectively show or hide each effect. The more experience you gain with Photoshop, the more layers you'll use. This next example shows a small fraction of the overall layers used to create the portrait, but you can see I used the layers to separately adjust the leaves, to change the color of her makeup, and to remove some shadows.

When you're using the Photography workspace, the Layers panel will be visible in the lower-right corner. By default, you'll only have a single Background layer. Create a new layer by clicking the **Create a new layer** button (shown next), selecting **Layer | New | Layer**, or pressing **Ctrl+Shift+N**.

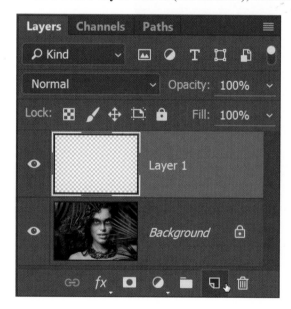

Now, select the new layer and use your brush tool (described earlier in the chapter) to paint a drawing over the picture. Show and hide your new drawing by clicking the eye icon to the left of the layer, as shown next.

That's the magic of layers—you can easily turn them off. Though it's simple, it's also incredibly powerful. Because you can turn layers off, you can easily undo a change you made an hour ago without undoing everything you've done since. You could even save your image and edit it another day.

For simplicity, I'll often make changes to the background layer in this book. However, in the real world, every distinct change you make should be done within a unique layer. That will keep your edits organized and ensure all your changes are non-destructive.

Masks

Click the eye icon to hide an entire layer, or create a layer mask to hide only part of a layer. The **Add Layer Mask button** (shown next) creates a solid white layer mask hiding everything in your current selection. To hide everything *except* your current selection, hold down the **Alt/Opt** key while you add a new layer mask. If you don't have anything selected, the layer mask will be solid white or black.

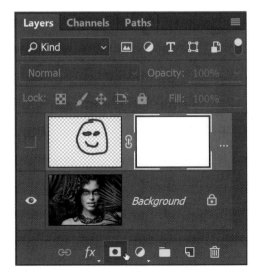

The white parts of a layer mask make that part of the layer completely visible. Black parts of a layer mask hide that part of the layer. Grey parts of the layer mask blend the layer with underlying layers, making those parts of the layer partially transparent.

In the next example, I selected the layer mask and used the Gradient Tool to paint a gradient from solid white to solid black, with shades of grey in between. Where the mask is white, at the top of the layer, Photoshop shows my drawing. Where it's black, at the bottom, Photoshop hides the drawing. The shades of grey cause the layer to smoothly fade into the background.

You don't need to look at the layer mask to paint in it. However, to see the layer mask, **Alt/Opt**-click it. You can delete a layer mask by selecting it and then pressing **Delete**.

Layer masks are incredibly useful for fine-tuning your changes and for blending together multiple pictures. Naturally, we'll cover layer masks in more depth later in this book.

Saving

You'll need to save your picture if you want to use it later. Press **Ctrl+S**, or select **File | Save**. If you close your picture, Photoshop will prompt you to save it.

If you opened the picture from Lightroom, Lightroom should automatically re-import it. If you don't see it, be sure **All Photographs** is selected in the **Catalog** pane, and turn off any filters. If you still don't find your updated picture, use a Text filter to search for the filename. If you still don't see it, try importing the saved file into Lightroom as if it were a new picture.

Photoshop will never overwrite raw images. If you were editing a raw image, Photoshop will save it as a new file with a .TIF extension by default.

TIF files include your layers and you can open them in Lightroom, so they're the best choice for most Photoshopped images, whether or not you use Lightroom or they were originally raw. If your picture is unusually large, Photoshop will warn you that it's too big for a TIF file, and you'll need to save it as a PSB file instead.

To share your picture online, first save your image as a TIF file so you can edit it later. Then, save a second copy as a JPG file that you can share.

To save a JPG copy, select **File | Export | Export As**, or press **Ctrl+Alt+Shift+W**. In the Export As dialog (shown next) select **JPG** for the format and set the quality to 60-80%. If you need the image to be a specific width or height, or you just need it to be smaller, use the Image Size panel.

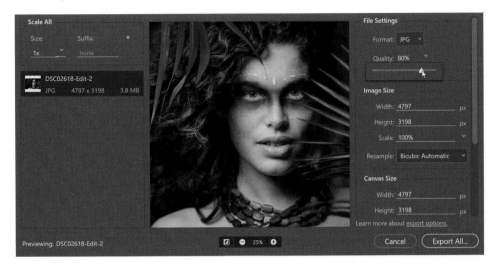

JPG is also the right format for sending pictures to online print services. JPG isn't the perfect file format, and it's very outdated, but high-quality settings still create excellent results.

That's it for Chapter 1! When you're ready to get serious, move on to Chapter 2 or watch some of our videos.

2 ADJUSTMENT LAYERS

Watch training videos at:
SDP.io/PSEDL

Use adjustment layers to make most minor photographic changes, especially fixing exposure and color problems and adjusting the saturation. Because each adjustment is a separate layer, you can edit or remove your adjustments after you create them, and you can use masks to apply adjustments to specific parts of a picture.

If you're using the Photography workspace, the Adjustments panel is shown by default along the right-center part of the window. If you don't see it, select **Window | Adjustments**.

When you click any of the adjustments, Photoshop adds a new layer above your currently selected layer. If you want Photoshop to rename it without any extra clicks, hold down the **Alt/Opt** key while you click the adjustment button.

When you create an adjustment layer, Photoshop displays the Properties panel (shown next). Use that panel to change the layer's settings. You can hide it again by clicking the >> at the top-right corner of the panel. If you want to change the adjustment layer later, double-click the layer in the **Layers** panel and Photoshop will again display the Properties panel.

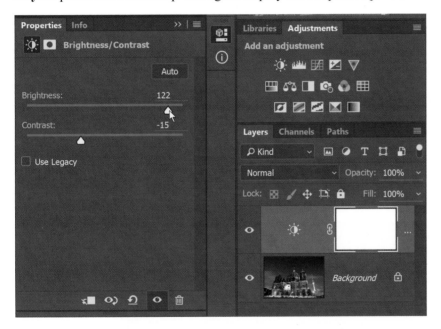

In this chapter, I'll show you how to use each of Photoshop's adjustment layers in the same order they appear in the Adjustments panel. Not every adjustment layer is useful for photographers, but I'll cover every adjustment layer type for completeness.

Before you skip ahead to a specific adjustment, you might read through the Brightness/Contrast section because there I'll show you the basics of working with adjustment layers and masks.

Brightness/Contrast

Like most night photos taken in brightly lit cities, this picture of Berlin has brightness and contrast problems. There are stars visible in the sky, but they're so dim that you wouldn't notice them. The archway and the tower in the background look solid white and overexposed, but because the photo was taken as a raw file, the camera captured detail in those parts that isn't currently visible.

The first step in solving these problems is to add the Brightness/Contrast adjustment layer by clicking the button, shown here:

When you click that button, Photoshop creates a new layer above our background layer. Photoshop also displays the Properties panel, as shown next.

I often start by clicking the **Auto** button in the upper-right corner of the Properties panel to have Photoshop take a guess at the best settings; it's often right.

You could adjust the **Brightness** and **Contrast** sliders to make the entire image look as nice as possible, but that would be more easily done in Lightroom. If you're making adjustments in Photoshop, it's probably because you want to apply your edits to specific parts of the photo. The easiest way to apply adjustment layers to specific parts of the picture is to select that part of the picture before creating the adjustment layer, so let's go back in time by pressing **Ctrl+Alt+Z** or **Cmd+Opt+Z** a few times (or just deleting the adjustment layer).

This time, I'll select just the sky before creating the adjustment layer. The easiest way to do that is to use the Quick Selection Tool, shown next:

With the Quick Selection Tool selected, I can drag my cursor across the sky, and Photoshop will select it automatically. If I miss part of the sky, I can drag across that part and Photoshop will add it to the selection. If I accidentally select part of the building, I can hold down the **Alt/Opt** key and drag in that part of the selection to de-select it. Eventually, I'll get the sky selected, as shown next. If it's not perfect, that's OK because we can fix it later.

One more selection step before I add the adjustment layer:

1. Click the **Select and Mask** button below the menu bar.
2. Expand **Edge Detection**, select **Smart Radius** and drag the **Radius** slider up until Photoshop does a good job of selecting just the sky.
3. Increase the **Feather** setting somewhere between 1 and 3 pixels so the edge is more natural.
4. Set **Output To** to **Selection**.
5. Click **OK**.

Now, my selection is pretty good, and I'm ready to add the Brightness/Contrast adjustment layer.
Because I had the background selected, Photoshop automatically selected the sky in the layer's
mask, as shown next. The adjustment layer will only change the parts of the picture that are white in
the mask, so my settings will change the sky, but not the building.

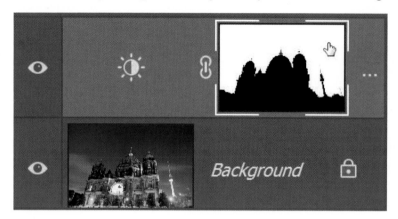

Now, I can adjust the **Brightness** and **Contrast** sliders to make the sky look great and show those
stars more clearly. If I applied those settings to the entire picture, the building would be completely
overexposed.

Tip: Double-click Brightness or Contrast to reset that setting. You can also click the Reset button on
the bottom-center of the Properties panel.

Now, I can reduce the brightness on the building by creating a second adjustment layer. First, I need
to select only the building:

1. If adding the adjustment layer cleared the selection, restore it using **Select | Reselect**.
2. Invert the selection by choosing **Select | Inverse.** pressing **Ctrl+Shift+I**, or pressing **Shift+F7**.

If I didn't have the sky selected, I could create a selection from the adjustment layer's mask. Press **Ctrl+A** or **Cmd+A** to select the entire picture, then right-click the layer mask and select **Subtract Mask From Selection** (shown next).

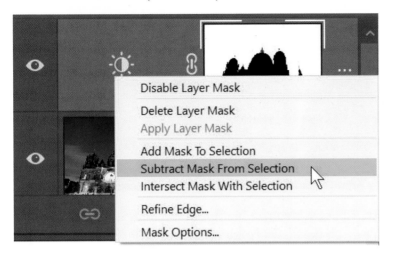

Either way, I now have the building selected, and clicking **Brightness/Contrast** will add a second adjustment layer, shown next. For this adjustment, I dropped the **Contrast** slider because the city lights were too harsh.

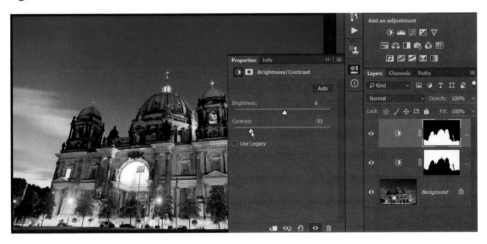

The archway in the building is still very overexposed, but I can fix that with a third Brightness/Contrast adjustment layer. First, I'll select the background layer. Then, to select only the brightest parts of the picture, I'll choose **Select | Color Range**. On the Color Range dialog, I'll set **Select** to **Highlights** and then adjust **Fuzziness** and **Range** until it shows just the overexposed parts of my picture (shown next).

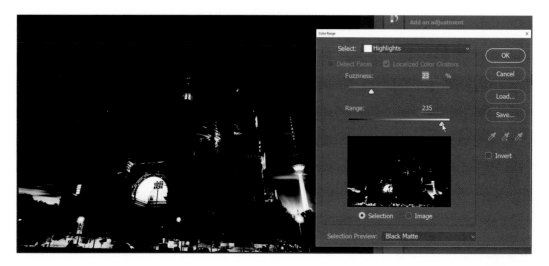

After clicking **OK**, I can create a third Brightness/Contrast adjustment layer that will apply only to the overexposed parts of the picture. Now, I can drag the **Brightness** down all the way to the left without changing the rest of the image. This shows the detail in the building's archway without ruining the picture.

There are a dozen different ways you could recover the stars in the sky and the overexposed highlights in the arch, and some might work better for you. I used this as an example to teach basic adjustment layer techniques, not because this is the only or best way to make these types of changes.

Levels

Use the Levels adjustment layer (shown next) if you already know what levels are and how to use them. If you aren't already familiar with levels, you'll probably find Brightness/Contrast and Exposure to be more intuitive. Remember to select the background layer before adding an adjustment layer.

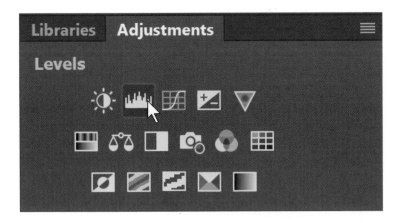

The Levels adjustment gives you direct control over the histogram. A histogram is a chart that shows the relative amounts of blacks, shadows, midtones, highlights, and whites in your picture. In the following example, you can tell that most of the picture is dark, because the histogram is higher in the left part of the chart. You can also see that parts of the picture are overexposed, because the histogram climbs up the right side.

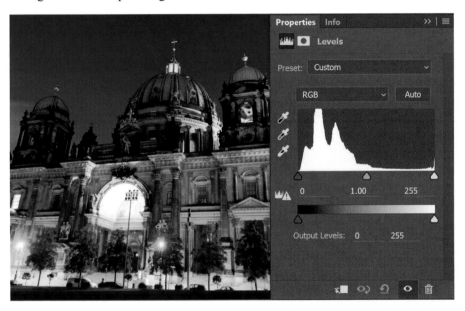

There are dozens of changes you can make using this panel; it's more complex than it might look. First, experiment with the black histogram slider. When you drag it to the right, you tell Photoshop that it should treat a specific shade of grey as black. As a result, all the shadows in the picture become black, hiding detail. The midtones now become shadows, and the brightest parts of the picture are mostly unaffected.

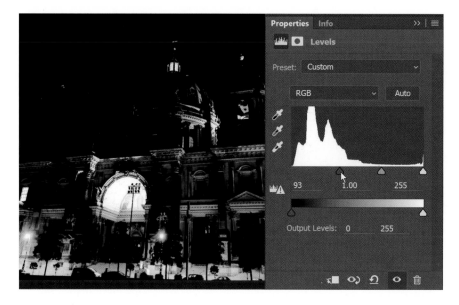

Next, experiment by dragging the white slider to the left. This tells Photoshop to treat the midtones as white, hiding all the detail in the highlights.

Raising the black slider hides detail in the shadows, while lowering the white slider hides detail in the highlights. The grey slider, in the middle, doesn't actually hide detail. However, it will compress the shadows (if you drag it to the left) or the highlights (if you drag it to the right). Effectively, dragging the grey slider to the left brightens the picture (shown next), while dragging it to the right darkens the picture.

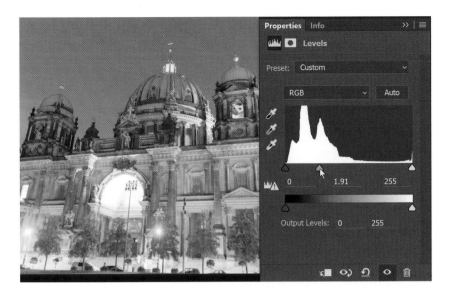

Besides manually dragging the black, grey, and white sliders, you can use the droppers to the left of the histogram to use part of the picture to define those levels. In the next example, I used the black dropper and clicked part of the building, telling Photoshop that I wanted the part of the picture I clicked to become pure black.

Below the histogram, you'll see the output levels. 0 is black, 127 is middle grey, and 255 is pure white. If you want Photoshop to use only middle grey to pure white, set the **Output** levels to 127 and 255, as shown next. This gives the photo a washed-out look, which isn't typically attractive but can be used for special effects.

When you change the output levels, Photoshop reassigns the image brightness. In the previous example, parts of the picture that were pure black would normally have a value of 0. However, because I have set the minimum output level to 127, Photoshop outputs 127 (middle grey) for everything in the image that is 0 (pure black). I didn't adjust the maximum output level, so 255 (pure white) still outputs as 255. Parts of the original image that are middle grey (127) will be output halfway between 127 and 255, with a value of 191.

Tip: Why is the number range 0-255 instead of 0-100? 255 is a round number to computers, which think in binary (0s and 1s). In binary, 00000000 is 0, while 11111111 is 255. If it were up to me, though, I'd show the values humans are more comfortable with.

We've only scratched the surface of what you can do with levels. Click the list above the histogram to separately adjust the levels for the red, blue, or green channels. Every dot in your picture is a single pixel, and every pixel consists of separate red, blue, and green values. If you mix different values of red, blue, and green, you can create any color combination.

The next figure shows circles filled with pure red, blue, and green, and the different color combinations each makes. Where all three circles overlap, you see white. Where none of three colors appear, you see black.

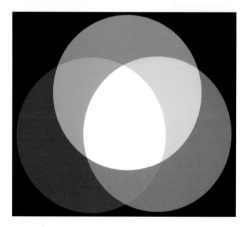

Unless you're color blind, your eye has three types of color receptors for red, blue, and green light. Camera sensors are designed exactly the same way. Similarly, typical monitors, TVs, and smartphones use red, blue, and green pixels packed closely together to simulate any color. As you can see, both digital imagery and human perception are based entirely on mixing these three primary colors. As a result, Photoshop includes some tools that help you manipulate the three channels separately.

In the next example, I selected the blue channel and then dragged the grey slider left, which made only the blues in the picture brighter. The effect was to brighten the blue sky and shift the color of the building from yellow-orange to more neutral.

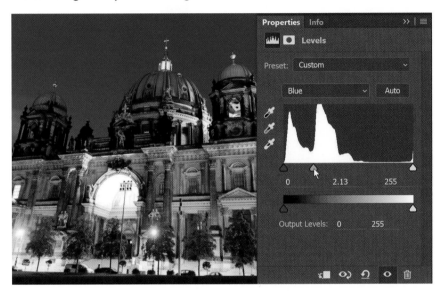

Want to see what the individual channels look like? Select Red, Blue, or Green from the drop-list, hold down the **Alt/Opt** key on your keyboard, and then drag the black or white slider. Photoshop will show you just the parts of the channel that are being clipped. The next channel shows all the blue shadows in the image because I held down the **Alt/Opt** key and moved the black slider.

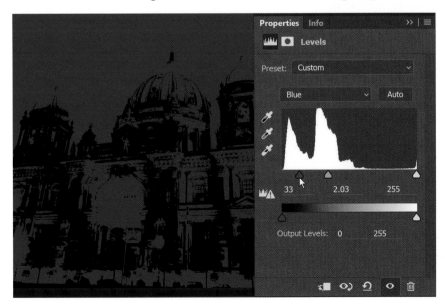

The levels adjustment layer has one more setting: the **Preset** drop-list. As you can see from the next example, you can use this list to quickly apply a setting. A more precise approach is to use the **Brightness/Contrast** adjustment layer for those settings and copy your image to a new layer and use **Image | Adjustments | Shadows/Highlights**.

Curves

The Curves adjustment layer (shown next) works very similarly to the Levels adjustment layer (but works more intuitively), so read the previous section if you haven't yet.

The Properties dialog for Curves shows, by default, a diagonal line running from the lower-left to the upper-right corner of the histogram. This line represents input vs. output brightness.

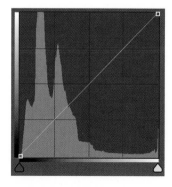

The left side of the histogram represents total black, while the right side represents pure white in your original photo. The bottom represents total black, while the top represents pure white in your edited picture. On the left and bottom edges of the histogram, you'll see a gradient from black to white that represents this relationship.

The default diagonal line indicates that an input value of 0 (pure black) will have an output value of 0. So, black parts of your original picture will be black in your edited picture. Input values of 255 (pure white) will have an output value of 255. So, white parts of your original picture will still be white in your edited picture. Similarly, middle grey (127) will output as middle grey.

You can grab that diagonal line and change the shape, altering the output values. For the next example, I dragged up the middle of the diagonal line. The left and right corners of the line stayed the same, so black remained black and white remained white. However, everything in between changed. Middle grey (127) will now output as 210, which is much brighter. The effect was to make the picture overall brighter, without changing the white or black points.

If I drag the middle of the line down, the entire image becomes darker (shown next). As you can see from the **Input** and **Output** labels below the histogram, middle grey values of 127 will output as the much darker 81.

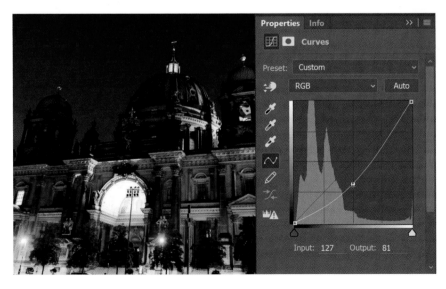

Remember, you can quickly reset an adjustment layer by clicking the appropriate button at the bottom of the Properties panel, as shown next.

You can add multiple points and curves to the line, but you won't usually get natural-looking results. In the next example, I added four points and reversed the white and black points, causing some of the shadows to be brighter than some mid-tones and highlights. The results are very strange.

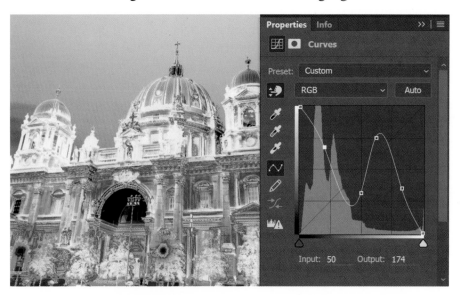

Just like the Levels adjustment layer, you can use droppers to set the white, black, and grey points. Use the hand icon (shown next) to specify a point on the curve by clicking a part of your picture.

While the results will never be natural, you can use the pencil icon to draw your own curve and create a colorful and abstract version of your photo.

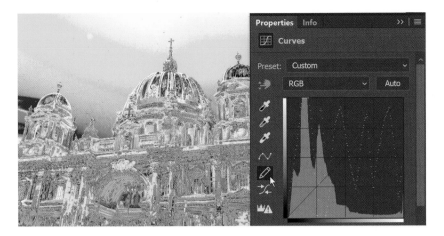

Photoshop provides a handful of useful presets. If you scan negatives and don't have software that automatically processes it, you can apply the **Color Negative (RGB)** preset to get more natural looking colors. The **Cross Process (RGB)** preset simulates an old darkroom processing technique that can create artistic tones.

As with the Levels adjustment layer, most beginning photo editors should avoid the Curves adjustment layer because of its complexity; results can look unnatural very easily. However, you can use the Curves adjustment layer to create specific special effects and to precisely solve some problems with shadow and highlight contrast.

Exposure

You can use the Exposure adjustment layer (shown next) to change the exposure of your picture, as if you had added exposure compensation when taking the photo.

Changing the exposure isn't the same as changing the brightness. When you adjust the brightness, Photoshop smoothly and intelligently alters the image to keep it natural looking. When you adjust the exposure, Photoshop unintelligently makes everything brighter or darker, potentially leaving you with washed-out highlights and crushed shadows.

You should never use the Exposure adjustment layer when working with a raw image. Instead, use **Brightness** because that tool can access the entire dynamic range of your raw file. In the following example, I lowered the exposure of the picture by 1 stop, hoping to recover some of the blown-out highlights. The left part of the image shows the results from the Brightness adjustment layer, while the right half shows the results from the Exposure adjustment layer. As you can see, the Exposure adjustment layer couldn't access the raw data from the overexposed parts of the picture, so whites became grey, and it revealed no additional detail.

In summary, don't use the Exposure adjustment layer.

Vibrance

Use the Vibrance adjustment layer (shown next) to brighten or fade the colors in your photo. Many people would instinctively use the Hue/Saturation adjustment layer for this purpose, but Vibrance is better for the same reason that Brightness is a better choice than Exposure.

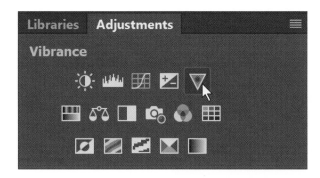

For this next example, I applied +100 Vibrance to the left half of the picture, and +100 Saturation to the right half. The Vibrance looks much more natural; the Saturation made the building orange.

The difference is that adding a lot of saturation will cause all bright colors to be equally saturated at 100%. Vibrance scales bright colors more smoothly.

Hue/Saturation

You can add a Hue/Saturation adjustment layer by clicking the button shown in the next example. You can use the tool to change a red dress to a green dress, add color to a black and white photo, or create spot color (also known as selective color) effects.

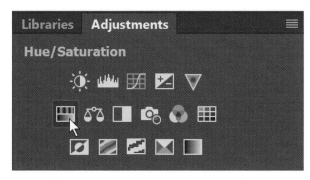

I discussed saturation in the previous section, so I'll focus on adjusting the hue. Drag the hue slider to the left or right to shift all the colors up or down the spectrum. The next example shows dragging the Hue slider to the left, which caused blues to shift to green, and yellows to shift to purple.

I've never changed the entire hue of a photo, however. This tool is more useful when you selectively change specific colors. For example, imagine you wanted to change the color of Aja's eye makeup. For the next example, I used a mask to apply the Hue/Saturation adjustment level to only the right half of her face so that you could see the original makeup on the left half of the image. I limited the range to the **Blues** because her makeup is blue and I didn't want to change all the colors, as shown next.

Notice that the color range selector at the bottom of the panel is now limited to the blue part of the spectrum. Next, I shifted the **Hue** slider until the blues became red.

That quickly changed the color of her makeup, but it's not perfect. Near the bottom of her makeup line, some of the blue is still visible because it fell outside the range of affected colors.

To fix that, I used the dropper tool to expand the impacted color range. By clicking the parts of the image that should be changed, the panel expanded the slider at the bottom. Unfortunately, it expanded it too far into the reds. While her makeup looks better, the color shift also impacted her lipstick, causing her lips to look yellow.

Rather than relying on Photoshop to automatically expand the range of affected colors, you can manually adjust the color range by dragging the outside edges of the slider at the bottom of the panel, as shown next. Dragging the right edge to the left caused the red tones of her lipstick to fall outside the affected range. You should always use the smallest color range possible to limited unwanted side effects.

Notice that both ends of the color range (where my cursor is pointed in the previous example) are darker grey. This darker grey area indicates that the effect becomes gradually less throughout the range, creating a more natural-looking adjustment. You can expand or reduce that range as needed.

The model also had some blue tones in the whites of her eyes that also became red, meaning I selected too much of the picture. To fix that, I painted black in the layer mask so the hue adjustment wouldn't impact her eye. I'll cover masks in more detail later in this book.

For the previous example, I clicked the list and selected **Blues**. A more precise approach is to use the color picker, shown next. With that tool, you can click to select a color range from your photo, click and drag to quickly modify the saturation of the selected color, or Ctrl-click and drag to quickly modify the hue of the selected color.

The Lightness slider at the bottom of the channel changes the brightness of the selected color range. By dragging it to the left, the model's makeup looks black.

As with other adjustment layers, this type includes a set of presets you can use to quickly create special effects. The next example demonstrates applying the **Sepia** preset to the right side of the picture.

Use the Colorize checkbox to add color to a black-and-white image. In the next example, I opened a black-and-white image and added the Hue/Saturation adjustment layer with a mask that applied it only to the top half of the image. I selected **Colorize** and then dragged the **Hue** slider to make grey parts of the image a subtle blue. Use the **Saturation** slider to add more or less color.

As an alternative, you can also use the **Black White** adjustment layer (discussed later in this chapter) to colorize a black-and-white photo.

Color Balance

The Color Balance adjustment layer, shown next, is probably not what you're looking for. If you want to fix the white balance on your picture (for photos with a strange color cast) you'll probably get better results using **Filters | Camera Raw Filter** or Adobe Lightroom.

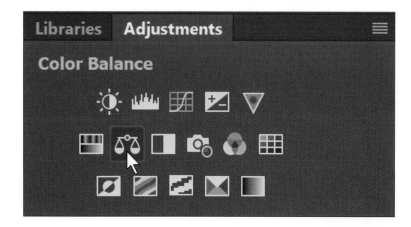

As discussed in the Levels section earlier in chapter, digital images are usually stored in three separate color channels. You can use the Color Balance adjustment layer to separately adjust the hue of each of these channels. For example, if you look at a picture and think, "The midtones have too much red and not enough blues," the Color Balance adjustment layer can fix that problem, as shown next.

Most humans don't look at a picture and mentally separate it into reds, blues, and greens, so don't feel bad if you don't think like that.

The labels on the Color Balance sliders are a bit confusing. It's easier to think of them as red, green, and blue sliders, and you're adjusting the relative levels of each. If you leave **Preserve Luminosity** selected, Photoshop will keep the overall brightness the same by raising or lowering the brightness of the other colors to compensate for any changes.

Use the **Tone** list to select whether you're modifying the balance of the shadows, midtones, or highlights.

Black White

Use the Black White adjustment layer to convert a color photo into black and white. You can also colorize a black-and-white photo, whether or not it started out as color.

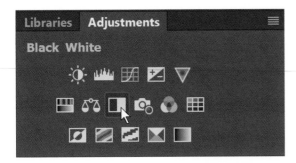

This adjustment layer is far more powerful than simply selecting **Image | Mode | Grayscale**. Converting a picture to grayscale simply removes all color information. The Black White adjustment layer gives you control over how bright each color is, simulating the colored filters that black-and-white film photographers used to selectively increase contrast.

By default, the Black White adjustment layer provides varying levels for six different color ranges (shown next). These levels are designed to create attractive black-and-white images from color photos, but most images will benefit from some tweaking.

For example, Chelsea is wearing a red dress in the original image. If I drag the **Reds** slider to the right, Photoshop will increase the brightness of all reds in the image, making her dress more prominent.

As shown next, if I drop the **Reds** slider to a negative number, she will be silhouetted.

Each color slider adjustment can profoundly impact the contrast in your final black and white image. There's no formula you can apply to the adjustments; I recommend simply playing with the sliders until you get the results that you like. Avoid having consecutive sliders at opposite ends, however, because that will create unnatural edges where gradual color transitions occur. In other words, be careful if you put the **Yellows** at -200 and the **Greens** at 300.

To colorize the image, select the **Tint** checkbox. By default, Photoshop adds a pleasant Sepia tint to the image. Click the color box to use the color picker to select any other color, as shown next. Click the vertical rainbow bar to change the color.

When using the Color Picker to add tint to an image, pick a color closer to the upper-right corner of the color picker to add more intensity. The further you move away from the upper-right corner, the subtler the color will be. The selection circle in the Color Picker box of the previous example is far away from the upper-right corner, making the colors fairly subdued. If I select the upper-right corner, as shown in the next example, the colors will be much more intense.

Adding tint to a black-and-white photo can give it a more artistic and faded appearance. Historically, many black-and-white printing types produced warm, sepia tones, hence Photoshop's default tint. I always experiment with adding a tiny amount of tint before publishing a black-and-white image online.

Photo Filter

Back in the film era, before we photographers regularly used Photoshop, it was common to attach colored filters to your lenses. These filters could either correct lighting problems (such as correcting the yellow color of incandescent lights when using daylight film) or they could add special effects, like adding a warm glow to skin tones in a portrait.

Save your money on the filters because you can create the same effect with Photoshop using the Photo Filter adjustment layer, shown next.

Typically, you'll use the adjustment layer by selecting one of the preset filters and then adjusting the density. In this next example, I use the Cooling Filter to add a blue tint to a photo taken in snow, making it visually feel colder.

These filters are one of the easier ways to correct white balance problems with Photoshop. If one of the existing filters isn't perfect for your image, you can click the **Color** box to change the color, and adjust the **Density** up or down to add more or less impact.

A real-world filter attached to the front of your lens will always block some portion of the light. Higher density filters might reduce the light by several stops. Unless you or your autoexposure system compensated for it, that would make your picture darker. By default, the **Preserve Luminosity** checkbox is selected, which will maintain the brightness of your image. If you clear the checkbox, your image will become darker.

Channel Mixer

Remember the discussion of red, blue, and green channels in the Levels section earlier in the chapter? If you really want to get crazy with these channels, use the Channel Mixer adjustment layer (shown next).

I've never used this for photography, but for completeness, I'll show you how it works. As mentioned above, each pixel in your image is composed of separate red, blue, and green input values. The channel mixer allows you to redirect one color channel to another color channel.

In other words, you can take all the red in your picture and make it green or blue. You could turn off all the blue in the image. You could double all the green.

In the Output Channel list, select the channel in your final edited image. As shown next, the defaults make sense: when you select the Green output channel, you'll see that Photoshop sends 100% of the green input channel to it, and none of the reds or blues.

So, by default, green is only green, red is only red, and blue is only blue. You could use the channel mixer to make the blue output channel use the red input values, and vice versa. I don't know why you'd ever do this for a photo, but that's what I did for the next example.

Color Lookup

Here's one of the most underused adjustment layers: Color Lookup. With this adjustment layer, you can make your pictures look like they were shot on film using hundreds of different profiles on the Internet.

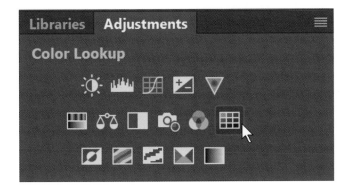

These looks are applied using look-up tables, or LUTs. These LUTs just map input values to output colors. In the next example, I applied the FoggyNight.3DL look to this book's cover photo.

Search the web for "film lut download" and you can find any number of free LUTs to create different effects. Most of these will be designed for video editing but they'll work fine in Photoshop; many filmmakers want to record with digital but have "the film look." The LUTs will have one of these file extensions: .CUBE, .3DL, .3DLS, .1DLS, .LOOK, or .CSP.

For the next variation on the same picture, I downloaded a free LUT pack from *sdp.io/smallhd* that simulates the looks of different popular films. I extracted the LUTs contained within the zip to a folder on my computer. Then, I added the Color Lookup adjustment layer, clicked the 3DLUT File list, selected the first option (Load 3D LUT), and browsed to the files I extracted. This look is cleverly named "Saving Private Damon."

You can apply looks with Photoshop, but my workflow saves adjustments like these for Lightroom. Lightroom makes it easier to flip through many different looks. You can also use Lightroom to create virtual copies of a single image, applying a different look to each, so that you can choose different versions for different uses.

Invert

The Invert adjustment layer (shown next) turns black into white, and vice-versa.

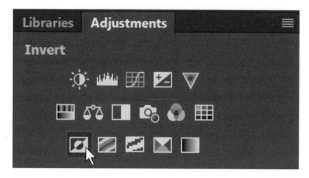

This adjustment layer has no options and doesn't have practical use for most photographers. If you need to process a color negative that you've scanned, use the Curves adjustment layer instead.

Posterize

Another almost-useless adjustment layer for photographers is Posterize (shown next).

Most images have 256 levels from black to white for each color channel. Posterize reduces the number of levels to the number you specify. As a result, instead of having smooth gradients between different shades, you'll see sharp edges and the colors will be off, as shown in the next sample.

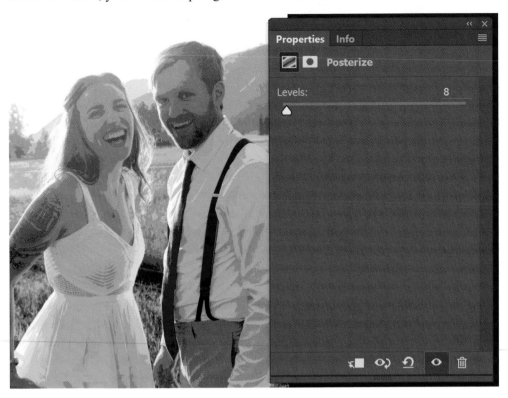

Threshold

Another infrequently used adjustment layer is Threshold, shown next.

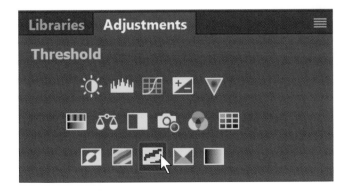

Threshold makes every pixel in your picture either black or white, dividing that level at the point you specify.

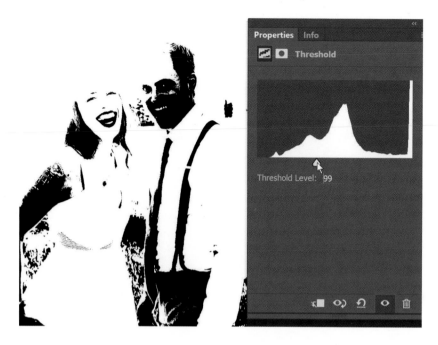

Selective Color

Use the Selective Color adjustment layer (shown next) if you've already tried the Hue/Saturation adjustment layer and thought, "I need way more control." Actually, you'll probably never need to use this adjustment layer, but I'll give you a quick overview and use it as an excuse to introduce a new color model.

With this adjustment layer, you select a color range and then control the levels of Cyan, Magenta, and Yellow, as well as the luminance (which is labeled Black). In the next example, I adjusted the yellow to make the people in the photo look more green by increasing the cyan and reducing the magenta.

The Selective Color adjustment layer is interesting because it uses Cyan, Magenta, Yellow, and Key/ Black (CMYK) instead of Red, Green, and Blue (RGB). CMYK, represented in the next figure, can be mixed to achieve any color, just like RGB discussed in the Levels section earlier in this chapter. It's just a different approach.

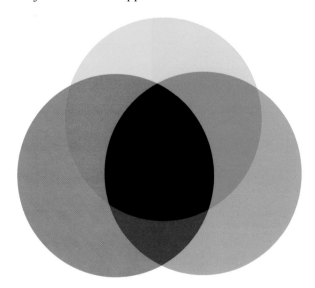

RGB works for mixing light. When there's no red, green, or blue present, there's darkness (black). When you mix them all, you see white; it's an additive model.

CMYK works for printing ink on white paper. When there's no cyan, magenta, or yellow present, there's white. When you mix them all, you see black; it's a subtractive model. That's why there's black in the middle of the CMYK wheels in the previous figure, but white in the middle of the RGB wheels in the Levels section.

Gradient Map

The Gradient Map adjustment layer (shown next) maps the varying brightness levels of your pictures to different colors.

For example, you can use this adjustment layer to make the shadows in your picture blue, the mid-tones red, and the highlights yellow, as shown next.

Photoshop includes many presets you can experiment with, and you can swap the highlights and shadows by selecting the Reverse check box. You can change the colors and adjust the transition points by clicking the gradient map box to open the Gradient Editor. In the next example, I adjusted the color midpoints (which appear when you move the center box) so that the color transitions added contrast to the most important parts of the same photo.

You'll probably never need to use the gradient map for photography, but it can create interesting effects.

3 HEALING, CLONING & PATCHING

Watch training videos at:
SDP.io/PSEDL

Have you ever seen a gorgeous landscape photo completely ruined by a powerline that you didn't notice when taking the picture? Have you ever taken a portrait that's perfect, except for the way the light exaggerated a blemish far more than anyone who met the person would notice?

With Photoshop, you can seamlessly remove distractions from your photos. Often, this creates a picture more like the scene you saw in your mind's eye. After all, when you're looking at a landscape in person, you don't fixate on the power line running through the scene, but you'd definitely notice it when looking at the still photo. If you were talking to someone, you wouldn't stare at a blemish on his or her nose, but our eyes are naturally drawn to flaws in a portrait.

These techniques can also be used creatively. You can turn a single goose flying through the sky into a full V formation. If you want to completely remove tourists from your photo of the Eiffel Tower, you can do that.

Photoshop provides several different tools for these purposes, and this chapter will discuss each in detail. Here's a quick overview of when you should use each:

- **Spot Healing Brush tool**. This is the usually the best choice; I almost always start with this tool, and I try the other tools only when it fails.
- **Healing Brush tool**. This is like the Spot Healing Brush tool, except that you must Alt/Opt-click to specify a source for the healing.
- **Patch tool**. Replace any area by dragging around it, rather than using a selection brush. Useful for removing the bags under eyes in a portrait.
- **Cloning**. Directly copies another part of the picture. Useful when a pattern must be perfect, such as removing a window from a brick wall, where the bricks must be perfectly aligned.
- **Content-Aware Healing**. Allows you to remove a distraction that you've already selected.
- **Content-Aware Fill**. Fills in any part of the picture using the same Content-Aware logic used by the healing tool.
- **Content-Aware Move tool**. Moves part of the picture, automatically filling in the empty spot.

Spot Healing Brush Tool

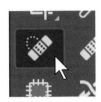

The Spot Healing Brush tool is perfect for quickly removing distractions and blemishes. As the name indicates, it works like a brush, so you can control the size and hardness of the edge. To heal something other than a spot, you can drag the brush.

I'll remove the orange tower in the background of this photo by dragging the Spot Healing Brush tool over it. The following examples show the photo before, during, and after the healing:

Most of the time, it really is that easy to use. Here are some general tips:

- Use a soft brush with a hardness of 20-50%. Adjust the brush hardness using the toolbar directly below the menus, as shown next.
- Use a brush large enough so that the center of the brush completely conceals the distraction. The edges of the brush are for blending, not for hiding.
- If you're removing a large distraction, use several passes and remove small portions of it with each pass.
- If you see weird artifacts after healing, click those artifacts again with the healing brush.

The toolbar (shown next) provides standard options for configuring the brush, and you normally don't have to adjust settings other than the brush size. The **Mode** should always be set to **Normal**.

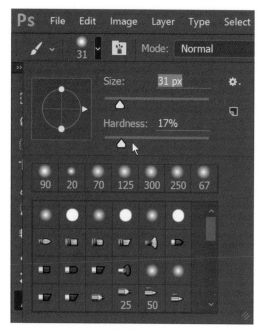

The **Type** should be set to **Content-Aware**, which is the most intelligent healing algorithm. If you're working with a multi-layer composite, you should usually select **Sample All Layers**.

The Spot Healing Brush tool does an amazing job of filling in the background when it removes a distraction. Adobe has put an incredible amount of engineering into this feature; the tool examines the background in an extremely intelligent way, imagines what would be behind the distraction you're removing, and fills it in by blending together other parts of the picture.

While it's usually quite magical, sometimes the Spot Healing Brush tool does a terrible job of removing a distraction. For example, consider this next before, during, and after example, where I attempted to remove a branch:

You, with your human brain, probably imagine smooth, green background blur behind that tree branch. Photoshop, with it's computer logic, imagined multiple cat faces were stitched together and hiding behind the branch. I could simply repeatedly re-heal the weird cat faces until Photoshop made less creative choices, or I could switch to a tool that gives me more control.

Healing Brush Tool

The Healing Brush Tool works almost exactly the same as the Spot Healing Brush tool, except that you specify the part of the picture to fill in the area you're healing. If you don't see it on the toolbar, select **Window | Workspace | Photography**.

To use the healing brush tool, first specify the source. Hold down the **Alt/Opt** key and click a good part of the picture—the source. This part of the picture will remain after you're done healing.

In this next example, I'm selecting part of the blurry background as the source in the first frame by holding down the Alt/Opt key and clicking. In the second frame, I'm dragging the tool over the object that I want to remove. The third frame shows the finished product, which is far better than what the Spot Healing Brush tool produced.

Some tips for using the Healing Brush tool:

- Use a soft brush. Hard brushes will create obvious lines.
- For large distractions, remove small sections at a time. If you remove something large, there's a good chance you'll move the mouse too far and you'll copy an unwanted part of your picture.
- If you do need to make multiple passes to remove a large distraction, you might need to occasionally select a new source point so you don't copy an unwanted part of the picture.
- Don't use the edges of the picture as the source; they won't blend well.

Clear the **Aligned** checkbox on the toolbar to always use the spot you originally clicked as your source. Select the **Aligned** checkbox to have Photoshop follow your cursor to change the source, even after you stop dragging the cursor.

Patch Tool

The Healing and Spot Healing Brush tools use brushes. If you'd rather select a distraction by dragging your cursor in a circle around it, use the Patch tool.

For example, I'll use the Patch tool to remove a distracting woman from the next photo.

That example would require at least one more pass to clean up the artifacts. I generally prefer to use the brush tools, but I find the Patch tool particularly effective at removing bags under eyes in a portrait.

On the Patch toolbar, note that the **Patch** option allows you to select either **Standard** or **Content-Aware**. Content-Aware is almost always the more effective healing algorithm.

Cloning

Cloning uses a brush to directly copy another part of the picture. In early versions of Photoshop, the Clone brush was the only tool photographers had to remove distractions, so the process of removing distractions became generically known as "cloning." Today, you'll often hear photographers use that term when they're actually using content-aware fill.

Cloning is very primitive, because it doesn't intelligently blend the copied area. However, it works better than the other healing options when repairing geometric patterns, such as bricks.

Imagine that you wanted to remove the blue paint drips from the bricks on the following photo:

It doesn't look too bad using the Spot Healing Brush tool, but if you look closely to the left of the cursor, you can see healing process disturbed the lines of the bricks.

To use cloning, first select the source by Alt/Opt-clicking. To make alignment as easy as possible, select a specific element in the pattern, such as the intersection of the bricks. Now, you'll see the pattern in your brush as you move your cursor.

You'll need to occasionally stop and reset your source so that you don't clone unwanted parts of the picture. Use a soft brush to make blending more natural.

Content-Aware Fill

If you'd rather select a distraction and then remove it, use Content-Aware Fill. After selecting the object to be removed, select **Edit | Fill**. In the Fill dialog, shown next, select Content-Aware. Then, click **OK**.

The next example shows the effectiveness of Content-Aware Fill at removing an object with a complex background. It's a less-than-perfect job, but a good start.

Copy, Paste, and Rotate

In the previous example of the blonde squirrel, I tested all the standard healing tools and none did a satisfactory job healing the corner of his eye, because too much of the eye was covered. There is plenty of eye and eyelid that could serve as a source for the corner of the eye, but direct cloning wouldn't allow it to match up.

Here's a trick: select another part of the eye and eyelid (make the selection bigger than you need), and then copy and paste by pressing **Ctrl+C** and then **Ctrl+V**. Photoshop will create a new layer for the copied section of the photo. You might not notice the new layer because it's exactly on top of the original layer.

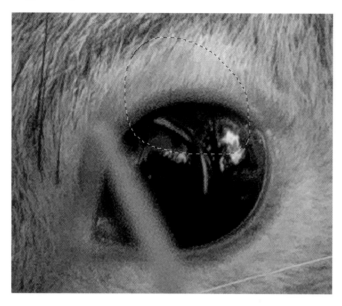

Select the Free Transform tool by selecting **Edit | Free Transform**. Move and rotate the copied piece of the eye so that the eyelid matches the curve, as shown next.

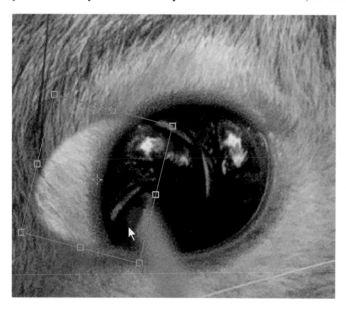

Press **Enter** to finalize the Transform. As discussed in Chapter 2, add a layer mask to the layer, and use a soft brush to paint black in the layer mask to smooth the edges and erase unwanted parts of the copied layer. After blending, the eye now looks like the following example.

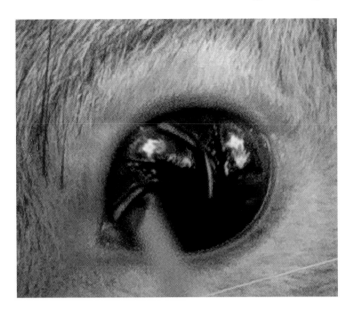

Repeat that process for the lower eyelid to fully recreate the eye, as shown next.

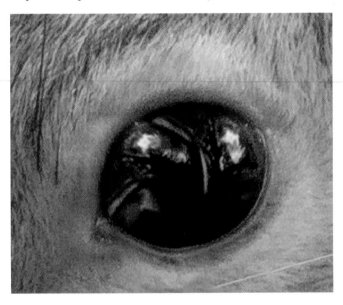

This technique takes some practice, but it's very useful for cloning shapes when the source material isn't perfectly aligned. If the curve isn't perfectly aligned, use the Liquify tool to smooth the edges.

Copy, Paste, and Flip

Copying and pasting is also useful for fixing errors in symmetrical buildings. For example, to remove the tree covering the left half of the White House in the following picture, I could copy the right half, flip it, align it, and then blend it into the original.

First, I'll select the right half of the building, copy it to a new layer (**Ctrl+C**, **Ctrl+V**), and flip it using **Edit | Transform | Flip Horizontal**.

Next, I'll temporarily set the opacity of the copied layer to about 50% so that I can see through it and better align it over the left half of the building.

I'll use the Free Transform tool (**Edit | Free Transform**) to move it and rotate it to match the left half of the building. I didn't need to rotate it, because the White House is quite level and I was perfectly centered, so the **Move** tool would have worked just as well. With the layer aligned, I can set the opacity of the layer back to 100%.

All that remains is to blend the edges and fill in the sky using the Spot Healing brush. It's done, and it only took about two minutes!

Content-Aware Move

Content-Aware Move moves part of the picture, exactly as if you had selected it and dragged it with the Move tool. After the move, however, Content-Aware Move fills in the empty spot.

In this next example, the owl is rudely violating the rule of space. Let's move him to the right half of the picture.

Before we move him, I'll use the Spot Healing brush to remove the branch. It doesn't have to be perfect because I'm moving the bird over it.

Now, I'll use the Content-Aware Move tool to drag the bird over.

Press **Enter** to finalize the changes, as shown next.

Photoshop did a pretty bad job of generating a new background. Content-Aware tools can save hours of time, but when they don't work properly, you'll still need to go in and manually repair parts of the picture. Still, this is a good start.

4 SELECTIONS

Watch training videos at:
SDP.io/PSEDL

You'll use selections to specify parts of a picture that an adjustment should impact. With practice, you'll become skilled at making selections, and your changes will appear completely natural. Photographers who aren't skilled with selections reveal their edits with unnatural edges for every change they've made.

I'll do my best to teach you the basics of creating great selections. However, making complex selections is more than just a skill—it's a *career*. There are skilled artisans whose entire careers are isolating every strand of hair blowing in the wind. I can't teach you that; it takes years of practice. I will show you how to use the most important tools, however.

Tip: Selections almost always go hand in hand with layers and masks. You should almost never simply select part of your picture and use that selection to directly edit your background layer. The next chapter describes layers in more detail because I have to start somewhere.

Note: Adobe almost never removes old features, even when a new feature is much more useful. In this book, but particularly this chapter, I'm going to focus on the most useful selection tools and give you only a quick overview of those tools that are mostly outdated.

Select and Mask

Since the release of Photoshop 2015.5, the Select and Mask tool (**Select | Select and Mask)** replaces the Refine Edge tool. For most images, Select and Mask should be the first stop for creating a new selection. This tool is simply amazing—it can give you an almost-perfect selection in a matter of seconds.

If Select and Mask seems slow on your computer, you might be happier using the more traditional Quick Selection tool, described later in this chapter. Performance varies depending on the resolution of your image and your computer's processor and graphics processing unit (GPU).

Select and Mask has many options, but most people will only need to know these tips:

- Use the **Quick Selection Tool** (the first tool) to intelligently select parts of your image.

- Use the **Lasso Tool** (the fourth tool) to manually select parts of your image that Quick Selection doesn't work well for.

- Click the **View** list and choose an option that makes it easy to see your selection.

- Under **Edge Detection**, select the **Smart Radius** checkbox and drag the **Radius** slider up until you get the best results. Often, I see the best results with high settings over 50 pixels.

- Under **Global Refinements**, use a **Feather** setting of 0.5-3 pixels, depending on how sharp the edges of your object are in the photo.

- Use a negative **Shift Edge** value (like -5%) to prevent the background from showing through.

- If you're using a green screen, Chroma Key, or other brightly colored background, select the **Decontaminate Colors** checkbox.

- Under **Output Settings**, in the **Output S** list, select **New Layer with Layer Mask**. This allows you to adjust the selection manually.

With experience, you'll learn which values work best. There's simply no substitute for practice and experimentation. No matter how long you spend studying the inner workings of the Select and Mask Tool, you'll often get the best results just by sliding different settings up and down to see what works with each new picture.

Tip: You can accomplish some of the same adjustments from the Select | Modify menu. However, I find the Select and Mask Tool much easier and more precise.

However, for those who want to understand every option on this tool, I'll list and describe them:

View Mode

Choose the option that makes it as easy as possible to see the accuracy of your selection. Depending on the selected view mode, you might be able to adjust the **Opacity** or **Transparency** up and down as needed. Often, showing it on black or white will make it easier to spot selection flaws. If the tool seems slow, try selecting **Marching Ants**.

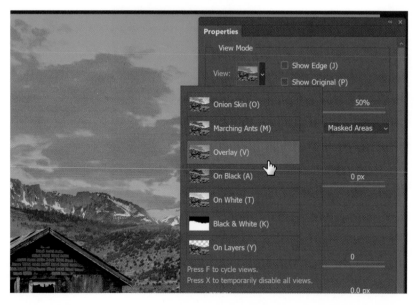

Show Original

Select this checkbox to temporarily view the selection as it was before Edge Detection applied its changes. Often, I'll select the checkbox to verify that my changes are improving the selection. Then, I'll clear it.

Edge Detection

The Edge Detection feature tells Photoshop to search inward and outward from your current selection, expanding or contracting your selection to match areas of contrast around your subject. The **Radius** tells Photoshop how far to search from your current selection. Drag the **Radius** slider around a bit and find the spot where the selection looks the best; it's more art than science. Change the **View** to better visualize the current selection.

Generally, I leave the **Smart Radius** checkbox selected. However, I always experiment with turning it off. Sometimes, the new selection is better with **Smart Radius** disabled.

Smooth

Selections can be a bit jagged and rough. Selecting a higher value for this slider smoothes out that jaggedness. Usually, a value from 2-5 works best. Zoom in tight on your image to see the difference.

This example shows the selection with **Smooth** set to 0. You can see the rough patches at the edge:

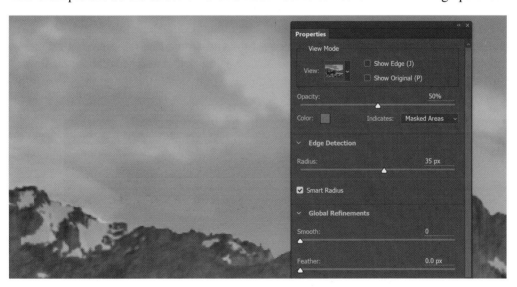

With **Smooth** set to 42, the selections are, for lack of a better word, smoother.

As with the other settings in the dialog, zoom in close to your selection and experiment with different **Smooth** values after you've chosen the best **Edge Detection** settings.

Show Edge

The **Show Edge** checkbox is only useful when you're using Edge Detection. To avoid confusion, Adobe really should move this checkbox to the Edge Detection section.

Select the **Show Edge** checkbox to temporarily view the range that Edge Detection is examining at the edge of your selection. Ideally, the radius would cover the entire area that the selection might need to be expanded or contracted. For example, if you were using Select and Mask to better select fine hairs against a background, the radius should be wide enough to cover the range from outermost parts of the hairs to the solid part of the head.

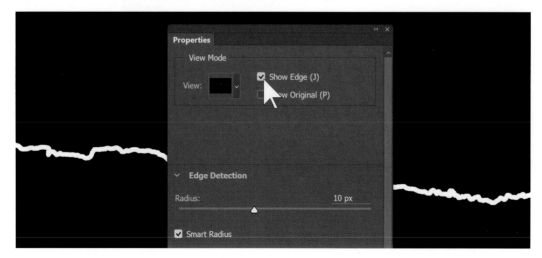

Adjust the **Radius** slider to increase or decrease the radius.

Feather

Without feathering, every pixel is either 100% selected or 100% deselected. That would lead to very unnatural blending because in photography all parts of a picture are a little bit less than 100% sharp, due to the effect of your sensor's anti-aliasing filter (if it has one), lens unsharpness, and blurring caused by parts of a subject being outside the focal plane.

In other words, foreground and background subjects naturally have a bit of feathering. If you copy-and-paste an object into a picture and don't use feathering, it won't look natural.

There's no one right answer to how much you should feather; it depends on how sharp the edges of the subject are on your picture. Again, the right answer is experimentation. With a sharp lens on a low-megapixel picture, 0.3 pixels might be the best setting. With an unsharp lens (or an out-of-focus part of the subject) on a high-megapixel picture, 5 pixels might be the best setting.

The next three examples demonstrate the selection (using the Black & White view) with **Feather** set to 0 pixels, 5 pixels, and 20 pixels.

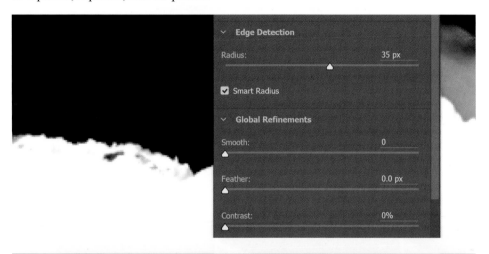

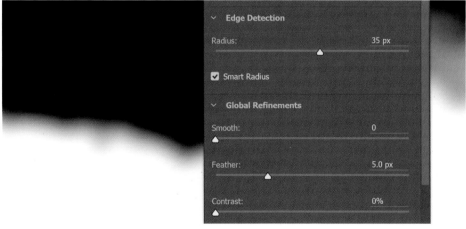

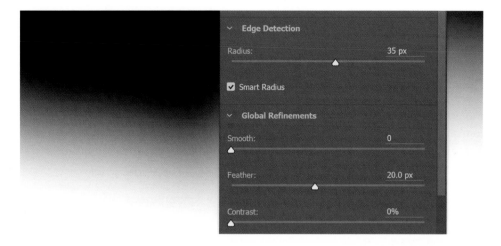

You don't necessarily need to use any feathering if you use Edge Detection. Often, Edge Detection will automatically handle the feathering.

Contrast

The **Contrast** slider works a bit opposite of the **Feather** slider. Rather than smoothing the selection to use grey values, raising the Contrast value eliminates grey values, hardening the edge of the selection.

The next three examples demonstrate the selection (using the Black & White view) with **Contrast** set to 0%, 30%, and 100%.

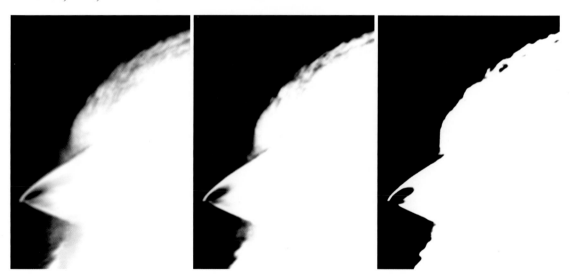

Shift Edge

Use **Shift Edge** to expand or contract your selection. If you see your selection includes a bit of the background, use a negative **Shift Edge** value until the background is completely removed. If you see your selection is clipping the edge of your subject, use a positive **Shift Edge** value.

The next three examples demonstrate the selection (using the Black & White view) with **Shift Edge** set to -50%, 0%, and 50%.

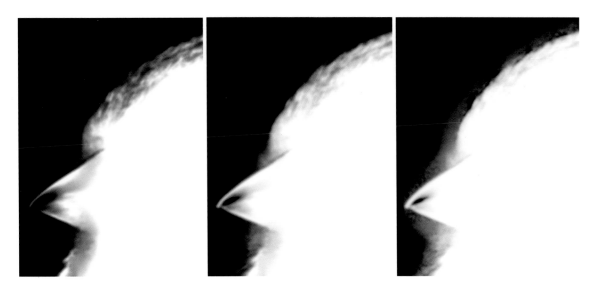

Decontaminate Colors

In the Feather section, I discussed how the edges of your foreground subject always slightly blur into the background. Feathering alone won't completely remove the background effect if the background is a significantly different color.

For example, the next two photos show a red fox on green grass. The first shows the original photo, the second shows the best selection I could get with Edge Detection, and the third shows the results with **Decontaminate Colors** selected.

When you compare the second and third picture, you can see that the third picture has much less green; that's the effect of **Decontaminate Colors**. If you were to move the fox to a non-green background, the green reflection from the original picture would look very strange.

Output To

The **Output To** droplist changes what Select and Mask does with your selection. The options, shown next, are self-explanatory. I almost always choose **New Layer with Layer Mask**.

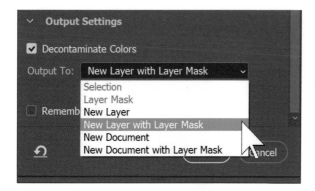

TIP: Ctrl/Cmd-click a layer mask to create a selection from it. Press **Ctrl+Shift+I** or **Cmd+Shift+I** to invert your selection.

Quick Selection

Since Photoshop CC 2015.5 introduced Select and Mask, you no longer need to use the standard Quick Selection tool. However, it still exists for times when the extra clicks to use Select and Mask aren't worth the trouble.

After selecting the Quick Selection Tool (shown next), simply drag inside the object you want to select, and Photoshop tries to find the edges of the object. If it doesn't select enough, simply click and drag again to add the rest of the object. If it selects too much, hold down the **Alt/Opt** key, and drag to remove that part of the selection.

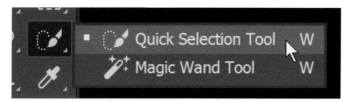

I used the Quick Selection tool to select the Junco in the following example, and it only took about five seconds.

If you look closely at the previous example, you'll see that it's good, but not perfect. After my initial selection, I always click the **Select and Mask** button on the toolbar, shown next.

Focus Area

If you want to select the in-focus or out-of-focus parts of your picture, use **Select | Focus Area**. It will automatically attempt to select the subject you focused on. Add more parts to the picture by **Ctrl/Cmd**-clicking them, or remove parts of the picture by **Alt/Opt**-clicking them.

As with the other selection tools, using **Select and Mask** almost always improves upon your initial selection. I find the default settings work best.

Color Range and Shadows/Midtones/Highlights

If your foreground or background is mostly a single color, you can quickly select one or the other using the Color Range tool (**Select | Color Range**). As shown next, the tool allowed me to separate the fox from the green grass in the background.

Notice that the **Invert** checkbox is selected in the previous example; because the background is all green, it's easier for me to select the background than the fox.

After opening the Color Range dialog, click the color in your picture that you want to select. To add more color range to your selection, hold down the **Ctrl/Cmd** key to add other colors to your selection. Adjust the **Fuzziness** slider up or down to refine your selection.

Click the **Select** list to choose a pre-defined range of color. This isn't as useful as you might think because colors in nature are never pure. For example, the grass in this picture is about as perfectly green as you can get, but selecting the **Greens** option ignores the cyans and the yellows that naturally appear in grasses.

The Color Range tool isn't that useful for selecting colors, but it is good at selecting highlights, midtones, and shadows. That makes it perfect for isolating darker subjects from a bright background or bright subjects on a darker background.

In this next example, my daughter's face is much brighter than the rest of the image. Therefore, I could easily select most of her face by using the Color Range tool, choosing **Highlights**, and then adjusting the **Fuzziness** slider.

Of course, there are always several ways to accomplish the same thing in Photoshop. The Quick Selection tool actually did a better job of selecting her face, even though I chose that image specifically as an example of how to use this tool.

Lasso

The Lasso is the most basic selection tool; simply drag your cursor to select a region.

The toolbar gives you options for feathering, as discussed earlier, and anti-aliasing, which should usually be selected.

I almost never use the Lasso tool because it's just not very precise. Occasionally, however, I'll need to add or subtract part of a selection, and Quick Selection isn't working properly. At those times, Lasso reliably does the job.

Magic Wand

The Magic Wand tool selects an area of similar color and brightness. In earlier versions of Photoshop, Magic Wand was the fastest way to automatically select parts of a picture. Ever since the introduction of the Quick Selection tool, I almost never use Magic Wand.

The toolbar (shown next) is very important to using the Magic Wand tool. Adjust the Tolerance value higher to select a wider range, or set it lower if Magic Wand selects too much of your photo.

You'll almost always want to leave the **Anti-alias** checkbox selected; it blends pixels more intelligently.

You'll usually want to leave the **Contiguous** checkbox cleared. When it's not selected, Magic Wand will only select parts of the picture that touch the area you clicked. When **Contiguous** is selected, Magic Wand will select matching parts of your picture, even if they're not connected to where you clicked. Select **Contiguous** when you want to select areas of sky between leaves, for example.

As with the Quick Selection tool, you'll get better results using the Select and Mask Tool afterwards.

In the next example, I used Magic Wand to select the background with a **Tolerance** of 10. It did a decent job but missed some patches of the background that were a lighter color. I would simply need to do more clicking to select those regions, too.

Magnetic Lasso Tool

Use the Magnetic Lasso tool to select any object that has visual contrast. Simply drag your cursor around the object, and Photoshop will attempt to find the edges and select it. As you can see from the following example, it's less than perfect, but it isn't restricted to straight lines.

For best results, zoom in tight on the subject you're selecting, and go slow. The tool will automatically add points during your selection, but if there's a sharp corner, click your mouse to add a selection point. To remove the last selection point, press **Delete**.

It depends on how steady your hand is, but for me, the Magnetic Lasso tool produces a much better selection on the first pass than the standard Lasso tool. However, I personally find the Quick Selection tool to be easier to use. For that reason, I usually start with the Quick Selection tool and only switch to the Magnetic Lasso tool if Quick Selection fails.

Selecting Geometric Shapes

Photographers capture the real world with their cameras, and the real world is rarely perfectly square or round. As a result, designers use the geometric selection tools more than photographers.

However, there are times when these tools are the most precise way to select an object. Windows, doors, and buildings are often perfectly rectangular. A full moon is pretty close to a perfect circle.

Rectangular Marquee Tool

Use the Rectangular Marquee tool in those rare circumstances when you need to select a perfect rectangle.

While most doors, windows, and buildings are rectangular, you rarely photograph them perfectly square. If you're standing at an angle to the object, if you're tilting your camera up or down, or if your lens has any distortion, the object won't be rectangular in your picture. The Polygonal Lasso Tool or Magnetic Lasso Tool might be better choices.

To illustrate this, imagine that you wanted to select the doorframe in the following picture. While the door itself is square, Chelsea was standing at a bit of an angle when she took the photo. As a result, the rectangle doesn't perfectly select the door.

Hold the **Alt/Opt** key while dragging to start your selection from the center, rather than a corner.

Elliptical Marquee Tool

Use the Elliptical Marquee tool to select oval or round shapes. Simply click-and-drag to select an oval. Hold down the **Shift** key to select a perfect circle.

Choosing where to start your click-and-drag is difficult. By default, Photoshop treats the Elliptical Marquee tool just like the Rectangular Marquee tool; wherever you click becomes one corner of your rectangular selection. That makes perfect sense with the Rectangular Marquee tool; you simply click on the corner of the object you want to select and then drag.

With the Elliptical Marquee tool, you need to imagine the circle or oval has a rectangle drawn around it and start dragging by clicking one corner of the rectangle. Alternatively, you could hold down the Alt/Opt key and start dragging from the exact center of the shape. Neither is very exact; it would be a much more useful tool if you could click the edge of your shape.

If you've photographed something round but shot it at an angle, it will appear elliptical and can still be selected with this tool. For example, the opening in the Samuel Beckett bridge in Dublin, Ireland is perfectly round. However, when I photographed it, I was standing at a bit of an angle. Therefore, my perspective makes it elliptical.

I did my best to select it, but because I'm human, I had a difficult time estimating the exact corner of the imaginary bounding box, and I didn't exactly find the center. You can drag and move your selection afterwards, but even then, the shape didn't exactly match. As a result, I found it easier to use the Quick Selection tool (described later in this chapter) to select the circle.

Polygonal Lasso Tool

The Polygonal Lasso tool selects any shape made up of straight lines. If there's a doorway and you photographed it slightly off-angle, use the Polygonal Lasso tool to select it by clicking each of the corners.

Here are a few tricks for using the Polygonal Lasso Tool:

- Hold the **Alt/Opt** key to draw curves.
- Press **Delete** to remove the last corner.
- Double-click to automatically complete the selection.

In addition to doorways, the Polygonal Lasso tool is perfect for selecting windows, buildings, books, or just about anything man-made with straight lines.

Creating a Layer Mask from a Selection

You'll often need to return to a selection you made in the past. The easiest way to do that is to make a new layer by using the Select and Mask tool (**Select | Select and Mask**) and selecting **Output To: New Layer with Layer Mask**. You can use Select and Mask even if you don't alter the selection, but I almost always use it to tweak the selection.

When every change you make is a layer, you can adjust the layer mask at any time. That's incredibly powerful because it allows you to expand or contract a change made earlier in the editing process.

Tip: Technically, you can load and save selections (**Select | Save Selection** and **Select | Load Selection**). In practice, it's usually easier to create new layers and use layer masks to define selections.

5 LAYERS

Watch training videos at:
SDP.io/PSEDL

Layers are, perhaps, the most important concept in Photoshop. Using layers, you can easily undo or modify changes that you've made at any point of the editing process. You can seamlessly blend together multiple photos. You can add text.

In fact, it's easy to end up with so many layers that they become disorganized. You'll have dozens of layers, and you won't be able to find that one layer you need to edit. Layers might get out of order, screwing up your entire photo.

In this chapter, I'll walk you through the most important concepts of layers for photography, including organizing your layers. We'll discuss using masks in the next chapter.

Locking and Unlocking Layers

When you open a picture in Photoshop, the software creates a partially locked background layer by default, as shown next. You can't move locked layers, but you can draw on them or otherwise screw them up.

You can lock any layer by selecting the layer and choosing **Layers | Lock Layers**. As shown next, Photoshop gives you the option to prevent changes to several aspects of it, but most photographers can simply choose **All**.

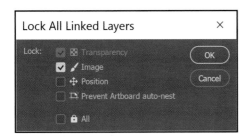

Locking layers that you don't need to edit further is good practice. When you create a complex image, which might have dozens of layers, it's easy to accidentally make edits with the wrong layer selected. Locking a layer reduces that risk.

If you want to unlock a layer, simply rclick the lock icon.

Adding Layers

However, if you keep the original picture, you can more easily refer to it or use it for healing and cloning. If you prefer that approach, start every editing session by duplicating the background layer using **Layer | Duplicate Layer**.

If you don't see the Layers panel, select **Window | Layers**.

Photoshop often automatically adds layers to your picture. For example, Photoshop simply can't add text to your background layer. If you use the Type tool, Photoshop will create a new layer, as shown next.

You can duplicate layers by right-clicking them and selecting **Duplicate Layer**.

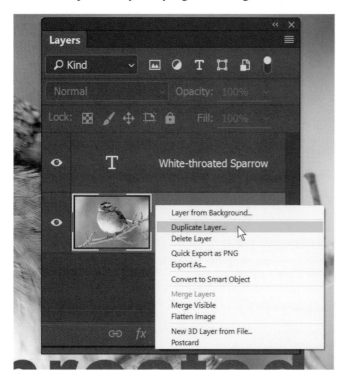

After you add a layer, double-click it to rename it. Naming layers doesn't seem important when you have only two or three. When you have five or more layers, though, logically named layers will make the editing much faster.

You can create a new blank layer by pressing **Ctrl+Shift+N**, selecting **Layer | New | Layer**, or clicking the New Layer icon on the Layers panel (shown next).

As a photographer, blank layers are the best way to add special effects like makeup or dodging and burning. I'll cover those effects in later chapters. As discussed in the previous chapter, the easiest way to make a new layer from a selection is to use the Select and Mask tool.

Opening Multiple Files as Layers

Photographers often want to open multiple pictures as layers. For example, if you're blending two images together into a composite, you'll need to open the images as different layers.

If you already have the pictures open separately in Photoshop, you can copy the first picture, switch to the second picture, and then paste the first picture as a new layer. Alternatively, you could use the Move tool to drag one image into another window.

If you haven't yet opened the pictures, select **Files | Scripts | Load Files into Stack**. As shown next, Photoshop prompts you to select the files that you want to open. Click **Browse** to select the files you want to open. If you have files already open in Photoshop that you want to add to the layers, click **Add Open Files**. If the images are of the same scene (for example, if they're bracketed), select the **Attempt to Automatically Align Source Images** checkbox.

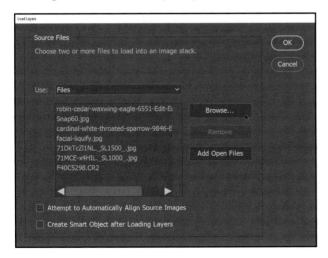

If you use Lightroom to organize your files, Ctrl-click or Cmd-click the files you want to open to select all of them. Then, select **Edit In | Open as Layers in Photoshop**.

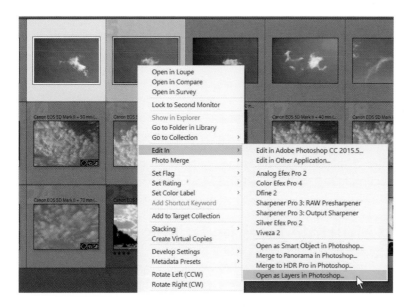

Merging and Flattening Layers

You can combine layers. Usually, it's best just to simply group multiple layers together. However, you might encounter filters or effects that work better after you've combined layers.

To combine layers, right-click the layers and select one of these three options, or select one of these options from the **Layers** menu:

- **Merge Layers**. First, select multiple layers by **Ctrl/Cmd**-clicking them. The layers will become one.
- **Merge Visible**. This option automatically combines all layers that aren't hidden.
- **Flatten Image**. This option automatically combines all layers, even if they're hidden.

After merging, you can't edit text or some other special effects that you previously created.

Moving, Resizing, Hiding, and Deleting Layers

Upper layers hide lower layers. For example, if I duplicate the background layer and move it on top of the text layer, it completely hides the text.

You can resize and move individual layers. Continuing the previous example, I used the Free Transform tool (**Ctrl+T** or **Edit | Free Transform**) to resize the top layer so that it didn't completely hide the lower layers.

You can make a layer partially transparent by adjusting the Opacity slider, as shown next.

To completely hide a layer without deleting it, click the layer's Eye icon. Hiding a layer is usually better than simply deleting it, because you might later change your mind and decide that you want to use that layer. If you delete it, you'll have to recreate the layer from scratch.

If you're sure that you'll never need a layer, click the layer and then click the trash can icon, as shown next.

Grouping Layers

Group multiple layers together for better organization. **Ctrl/Cmd**-click each layer, then select **Layer | Group Layers**.

The cover shot of this book has dozens of layers; the eye enhancements alone consist of three separate layers, shown next. Organizing them into groups and logically naming them by double-clicking the groups is a must.

You can move groups of layers together, making them almost as easy to work with as a single layer.

Linking Layers

You can link two or more layers so that Photoshop always moves them together. In other words, if you move one layer, Photoshop will move all linked layers to match.

To link layers, select the layers by Ctrl-clicking them. Right-click the layers, and then select **Link Layers**. As shown next, Photoshop will show the link icon beside the layers.

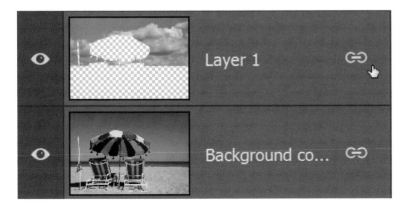

After linking those two layers, I moved the top layer. As you can see, Photoshop moved the linked layer to match, keeping the blended picture intact.

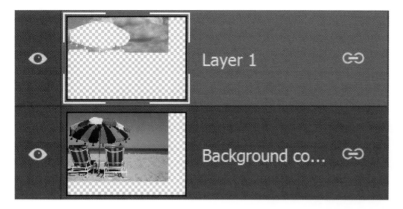

Finding Layers

It's easy to lose track of a specific layer when your photo gets complex. At the top of the Layers panel, Photoshop includes a tool for finding layers.

By default, this tool is set to filter layers by **Kind**. Selecting the different boxes displays or hides image layers, adjustment layers, text layers, shape layers, and smart objects. The next figure demonstrates filtering layers to show only image and text layers.

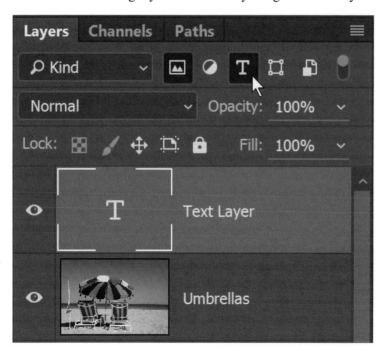

If you're wise enough to logically rename your layers, you can easily find them using the Name tool. As shown next, I had two layers that I had used for the Hard Mix blending option, and I renamed each of them. Searching for the word "hard" filtered the list to just those two layers.

After finding the layer, I selected it and turned off the search filter by clicking the switch in the upper-right corner (shown next).

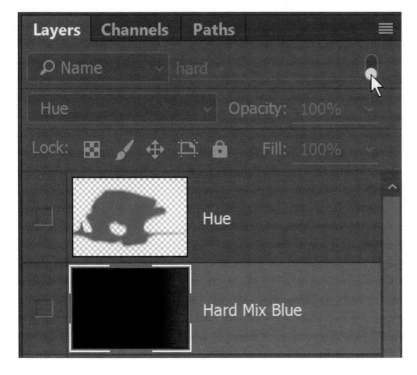

If you haven't renamed your layers, you can find them by searching for different properties. For example, select the **Mode** search tool and then select **Hue** to find all layers with the Hue blending mode.

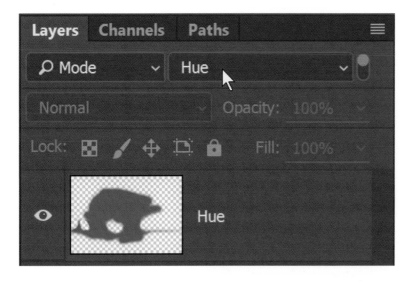

Coloring Layers

You can assign colors to your layers to help you find them more easily. This doesn't change how the layers function; it just helps you organize the layers.

To assign a color to a layer, right-click it and then select the color you want to assign. As shown next, colored layers are much easier to find.

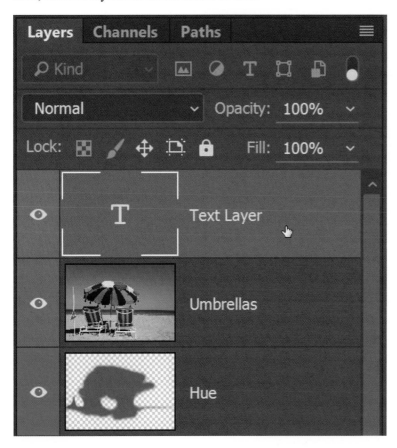

Blending Layers

By default, layers simply hide everything beneath them. You can use the Blending drop-list to choose from a wide variety of different blending options, each with its own purpose.

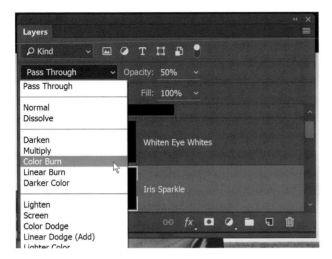

The sections that follow describe each blending option in more detail. It's impossible for me to cover every creative use for blending options, but I'll describe some scenarios that I've found particularly helpful.

Here are some quick tips:

- Use Screen to blend the moon or fireworks into a landscape.
- Use Overlay to add makeup to a photo.
- Use Soft Light to naturally add or remove fill light, or to add a color tint.

Normal

The default value works quite logically; upper layers hide underlying layers.

Lighten & Darken

The Darken blending option causes the layer to only make an image darker, never lighter. Lighten is just the opposite; it causes the layer to only make an image brighter, never darker.

The next three examples show a text layer over a photo background. This first screenshot has the text blending set to Normal, and as you can see, the text completely hides the underlying photo.

This second screenshot has the text blending set to Darken, which looks exactly the same on the bright parts of the photo. Darken allows dark parts of the underlying picture (such as the shadow on the branch) to show through.

With the blending set to Lighten, the text almost completely disappears. With Lighten, the upper layer can only make the picture lighter, and never darker. The text layer is quite dark, however, so you can only see it lightening the darkest parts of the picture.

Notice that Darken and Lighten are complete opposites. The parts of the text that are hidden with blending set to Darken are the only parts of the text that are visible with the blending set to Lighten.

So that's how Lighten and Darken work, but what are they good for? Lighten is perfect for adding a photo of a moon to a landscape. If there are clouds in the sky, or the sky isn't completely black, Lighten will blend the moon naturally into the sky. The next two examples show how it looks to blend the moon into the sky with Normal and Lighten settings.

Tip: White parts of a Darken layer disappear. Black parts of a lighten layer disappear.

Screen

The Screen blending option works similar to Lighten but generally produces more natural results. Screen is perfect for blending a photo of the moon or fireworks into a landscape.

The next example shows the moon blended with Screen and Lighten. As you can see, Screen looks more natural. Blurring the moon shot a bit would make it look even more natural.

Overlay

Overlay works very similarly to Screen. However, Screen only lightens the lower layer, while Overlay will lighten and darken the lower layer. This example shows the moon blended into the sky with the Overlay blending option:

Though it won't work on all backgrounds, Overlay blending can be useful for subtly integrating a watermark into your picture, as shown next:

Lighter Color & Darker Color

These blending options behave very similarly to Lighten and Darken. However, they tend not to blend as smoothly, so I don't recommend using them.

The next example compares the Darken and Darker Color blending options. Notice how Darker Color blends less naturally.

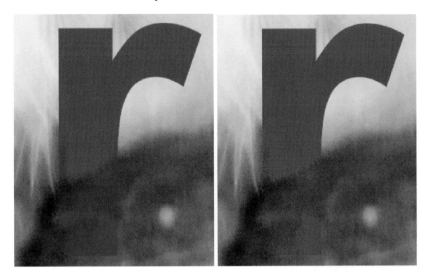

Dissolve

The Dissolve blending mode isn't particularly useful to photographers, but for completeness, I'll describe it. Dissolve randomly takes pixels from the edge of an object.

The next two screenshots show an extreme close-up of the top of a lowercase "r" blending into a mostly white background. The first screenshot has the Normal blending mode, while the second has the Dissolve blending mode.

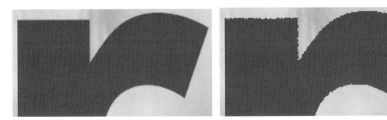

Multiply, Divide, Difference, Exclusion, Subtract, Linear Dodge (Add)

These blending options aren't particularly useful to photographers. As the names imply, Photoshop performs mathematical calculations for each individual pixel of the layers…but light doesn't multiply or divide in nature, so the results don't look natural.

Nonetheless, they can create interesting, creative special effects. Stack multiple pictures and play with these blending modes to create bizarre looks.

Color Burn & Color Dodge

First, remember that in Photoshop and darkroom processing, the term "Burn" means to darken, and "Dodge" means to lighten.

Color Burn and Color Dodge darken or lighten the lower layer by modifying the colors. Neither is frequently useful to photographers. Because I need a subject with color to demonstrate this, I'm going to create a very cheesy example. Please don't hold this against me.

This example shows the effect of Color Burn:

This example shows the effect of Color Dodge:

I've never used either blending option for photography, but these examples might be perfect for a design to be airbrushed onto a van from the 1970s.

Linear Burn & Linear Dodge

These produce results similar to Color Burn and Color Dodge, but the linear variants tend to look more natural. This example shows Linear Burn:

This example shows Linear Dodge:

Soft Light

This blending option darkens or lightens lower layers and is usually used by painting white or black on a new layer and then adjusting the opacity of the blending layer to make it look natural.

For example, I created a new layer and used the Brush tool to paint black and white. This next example shows that layer with the Normal blending option:

Now, I'll set that layer to Soft Light and set the opacity to 50%. The next two examples show the original image and the image with the Soft Light applied. As you can see, where I painted black, the image became darker in a very natural way. Where I painted white, Photoshop seemed to add fill light.

If you've ever wished you could go back and add more fill light, or block a bit of fill light, the Soft Light blending option can recreate the effect in post.

Soft Light is also an excellent way to add a colored gel to a light in post. For example, I clumsily painted blue over the model. After setting the blending option to Soft Light and setting the layer opacity to 70%, Photoshop added a gentle tint to the image.

Hard Light

Hard Light behaves like a more extreme version of Soft Light. Hard Light darkens or lightens the blended results while also increasing contrast. It's easier to demonstrate than to explain.

This example shows a layer of painted purple with the Normal and Hard Light blending options.

As you can see, the Hard Light blending option used the brightness of the underlying layer to determine the brightness of the mask.

The Hard Light blending mode is useful for creating a contrasty glow effect in photos. First, the original photo:

Next, I'll duplicate the background layer and set the blending mode for the new layer to Hard Light. As you can see, this greatly increases the photo's contrast:

Finally, I'll add a 100 pixel Gaussian Blur (**Filter | Blur | Guassian Blur**) to the top layer. I set the **Opacity** to 65% to tone down the effect. The image has a glow, but it's still very sharp. The final result is highly stylized, and certainly cheesy, but the mood is appropriately romantic for some uses.

Vivid Light

Vivid Light is a great way to add contrast to an image. Duplicate the layer, set the blending mode to Vivid Light, and then adjust the Opacity as desired. For this next before-and-after example, I set the Opacity to 18%.

That's the most common use for Vivid Light. However, to my eye, the results are indistinguishable from applying the Brightness/Contrast adjustment layer.

Linear Light

Like with Vivid Light, photographers often duplicate an image and apply the Linear Light blending mode to increase the contrast of the image. Again, however, my testing shows that applying the Brightness/Contrast adjustment layer creates similar results and is more efficient.

Pin Light

Pin Light is most often used to create special effects by painting colors over an image. For best results, the colors should not be 100% bright or dark, but somewhere in between.

In the following example, I painted pure 100% red on the far left, followed by 66% red, 33% red, and black. The brightest and darkest colors don't blend at all, but mid-tones will blend with the lower layer, adding their color.

Hard Mix

The Hard Mix is a special effect blending mode that makes your image two tones. Paint a color onto the blending layer, and the result will be either the purest version of that color or black. The color shouldn't be 100% bright; choose a color from the mid-tones. Anything brighter than the color you select will appear fully bright, and anything darker than your selected color will be black.

Consider these two layers:

They create this result:

Hue

Hue changes all colors to match the color you paint into the layer. In this next example, I used the Dropper tool to sample the blue color of the sky. Then, I hurriedly painted that blue color over the chairs, umbrella, and flag, and set the blending option to Hue.

 +  =

As you can see, where I painted blue, Photoshop changed any color in the underlying layer to match while retaining the luminance/brightness of the original image. There are many ways to change colors in Photoshop, but this is one of the easiest.

Color

Color works similarly to Hue, but adds color even where there is none. Here's the same example from the Hue section, but with the blending option changed to Color:

As you can see, the white parts of the umbrella and sand also changed color, making my sloppy brush work more obvious. Use the Color blending mode to add color to an image that doesn't have enough color, for example, when colorizing a picture or adding makeup to a model.

Saturation

The Saturation blending option replaces the saturation levels on the lower level with the saturation levels on the upper level. It's not especially useful for photography, but it can be used for special effects.

For example, if you paint colors with the maximum saturation on the blending layer, all colors in the lower layer will be 100% saturated.

First, I'll click the foreground color:

In the color picker, I'll click the upper-right corner to pick a color with maximum saturation. Then, I'll click **OK**.

I created a new layer and used the Brush tool to paint a bright blue stripe on the left side. I repeated the process to select a color from the middle of the color picker (so that it would be 50% saturated) and the bottom-left corner (so that it would be 0% saturated). Here's what the blending layer looked like while the blending option was still set to Normal:

Switching the blending option to Saturation produced this result:

Notice that the parts of the picture I painted with the maximum saturated color are now oversaturated. The parts of the picture that I painted with the mid-saturated color were brought to that middle level of saturation; the reds were hardly affected, while the white of the umbrella and beige of the sand had more saturation added. On the right, where I painted black, Photoshop completely desaturated the image, making that part of the picture black and white.

That's how the Saturation blending option works, and perhaps you can find a creative use for it. I find it more intuitive to use the Saturation adjustment layer with masks.

Luminosity

The Luminosity blending mode changes the brightness of a layer without changing the colors. Consider this example, where I layer a cloudy sky (that I colored pink) over a blue sky:

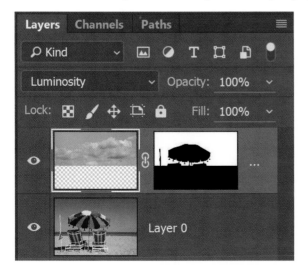

The blended result shows the sky with the same color as the original. However, the brightness of the top layer has been retained, so the clouds seem to appear in the sky.

Use the Luminosity blending option when you want to keep the original colors but add texture.

6 MASKING

Watch training videos at:
SDP.io/PSEDL

Use masks to show or hide parts of a layer. Masks allow you to seamlessly blend together two images.

There is another way to hide parts of a layer in Photoshop: the Eraser Tool. In theory, you could produce the exact same results using either masks or the Eraser, but masks have a huge advantage: they're reversible.

Once you delete part of a layer with the Eraser, it's gone forever. For that reason, I don't even teach you to use the Eraser. Whatever you can do with the Eraser, you can do better with masks.

With masks, it's easy to hide part of a layer by painting black in the layer mask. If you hide too much, switch your brush to white and paint the mask again to reveal parts of the layer. When you save your image as a .TIF, .PSD, or .PSB file, your layers and masks are saved, too, so you can make changes after closing Photoshop.

This chapter serves as a brief introduction on how to use masks. Be sure you understand these fundamentals before moving on because masks are essential to using Photoshop and this book.

How Masks Work

Masks attach to a layer in your image. They work as if they were overlaid on top of the image. Wherever the mask is black, that part of the layer is hidden.

To demonstrate this, let's start with a simple, abstract example. This image has two layers: the top layer is solid blue, and the bottom layer is solid red. Naturally, the red layer is completely hidden by the blue layer.

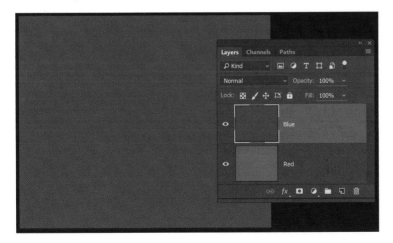

If we wanted to show the red layer, we could use the eraser tool and erase part of the top layer. Here's the word RED in my very best mouse handwriting.

What if I save the file, close Photoshop, and then want to change it to say RAD instead? Once I reopened the file, I wouldn't be able to undo my erasure, so I'd have to redo the entire project. That would waste several seconds of time on this project… but in a real project, you could easily lose hours of work.

I'll accomplish the same result using layer masks. First, I'll select the Blue layer and click the Add Layer Mask button:

Photoshop adds a solid white layer mask to the Blue layer:

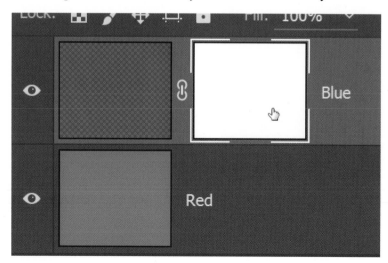

This doesn't change the picture. Because the layer mask is solid white, the entire blue layer continues to display as normal.

If I write in the mask using black, the lower layer will show through. First, I'll press **D** to set the foreground and background color back to white and black.

Now that my foreground color is black, I'll select the Brush tool by pressing **B**. Instead of painting on the blue layer itself, I'll select the layer mask. I won't see the layer mask directly, but I will be painting on it.

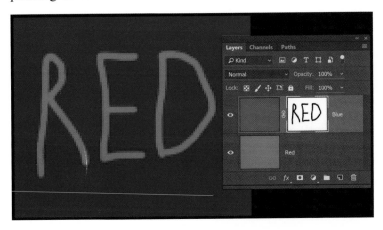

To see the layer mask directly, **Alt-** or **Opt**-click it.

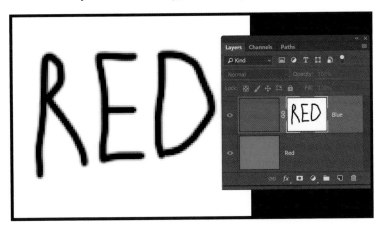

If I want to change the RED to RAD? I'll select the layer mask and then paint white over the E. An easy way to do this is to press **D** to reset my foreground and background colors to black and white and then press **X** to switch white to the foreground color. Then, press B to select the **Brush** tool. Force yourself to use these keyboard shortcuts; you'll be using them for years.

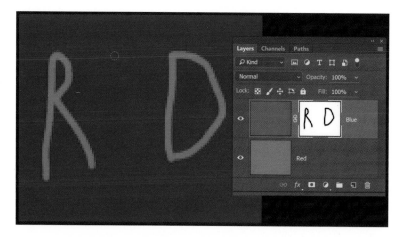

Now, I'll press X to switch Black to my foreground color, and paint the A on the layer mask. Easy!

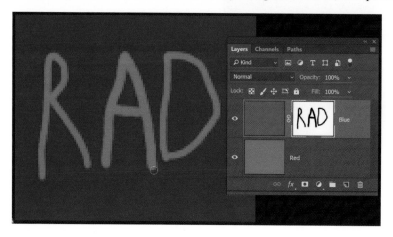

Masks don't have to be only black and white; shades of grey will show the lower layer with less opacity. For example, if I paint with grey, Photoshop will blend together the blue and red layers to form purple:

If I paint a white-to-black gradient in the mask, Photoshop will smoothly blend the images together. Later in this chapter, I'll demonstrate how to use gradients to blend two photos together.

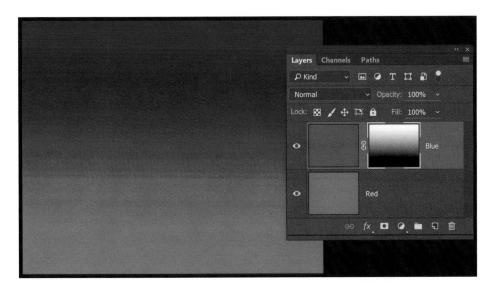

Basic Masking Tasks

You can view, move, copy, and apply masks, as well as accomplish many other useful tasks. Photoshop makes them all very simple, so I'll just list the capabilities and tell you where each is in the user interface:

- **Fill a mask with white or black**. Press **D** to reset your foreground and background colors. Select the mask, and then press **Alt-Del** (on a PC) or **Opt-Del** (on a Mac). To fill it with black, also press **Ctrl-I** or **Cmd-I**.

- **Delete a layer mask**. Right-click the mask and select **Delete Layer Mask**.

- **Move a layer mask**. Drag it with your mouse to a different layer.

- **Copy a layer mask**. **Alt**-drag (on a PC) or **Opt**-drag (on a Mac) the layer mask to another layer.

- **View or hide a mask**. **Alt**-click (on a PC) or **Opt**-click (on a Mac) the layer mask.

- **Invert a mask**. Select the layer mask and press **Ctrl+I** or **Cmd+D**.

- **Create a selection based on a mask**. Press **Ctrl+D** or **Cmd+D** (to clear your current selection) and then **Ctrl**-click the mask. Press **Ctrl+Shift+I** or **Cmd+Shift+I** to select the opposite mask.

- **Apply a mask.** Permanently delete hidden parts of your layer by right-clicking the mask and then selecting **Apply Layer Mask**.

- **Disable/hide a mask. Shift**-click the mask.

- **Refine the mask (to automatically improve a complex selection).** Right-click the mask and then select **Select and Mask**. Refer to Chapter 4 for information about the Select and Mask tool.

If you forget the keyboard shortcuts, right-click the mask to view a menu with most of these capabilities. Memorizing the shortcuts, though, will save you time in the long run.

Changing the Density of a Mask

You can change the density of a mask to show more of the hidden parts of the mask. It's the opposite of opacity, which hides more of a layer.

To change the density of a mask, double-click it to view the mask properties. Then, drag the **Density** slider down.

Masking Example: Adding Graffiti

Here's a practical example of masking: we'll combine these two photos to add graffiti to the wall behind the car.

1. Open the photos as layers, with the graffiti above the car photo. If you're opening the photos from Lightroom, select them both, right-click them, and then select **Edit In | Open as Layers in Photoshop**. If you're opening the photos directly in Photoshop, select **File | Scripts | Load Files into Stack**.

2. Set the **Opacity** of the top layer to 50% so that you can see through it to align it.

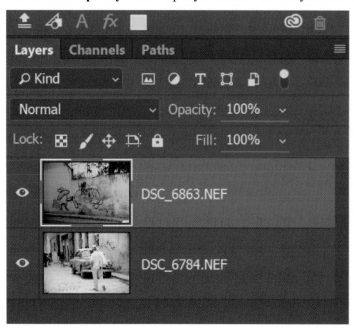

3. Use **Free Transform (Edit | Free Transform)** and **Perspective Transform (Edit | Transform | Perspective)** to size the graffiti to the wall and to match the angled perspective of the wall, as shown next.

4. Set the top layer **Opacity** to **80%**.

5. Set the Blending Option for the top layer to **Pin Light.** Or, choose whichever blending option makes the graffiti look most natural to you. I often scroll through all the blending options by clicking the list and then pressing the cursor down arrow until I find the option I'm most happy with. Your image should now resemble the next screenshot.

6. Now, create a white layer mask for the top layer. This won't yet change your picture.

7. Select the Brush tool by pressing **B**.

8. Set the paint color to black. If necessary, click the press **D** (to set the foreground and background colors to the default of black and white) and then press **X** (to switch black to the foreground color).

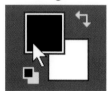

9. Select a big, soft brush. For example, set the **Size** to **350px** and the **Hardness** to **0**. The soft edge of the brush will blend the graffiti layer more naturally.

10.Use the Brush tool and paint black into the layer mask to hide the edges of the graffiti layer so that it blends naturally. If you paint too much, press **X** to switch the color to white and then repair your layer mask. To return to painting black, press **X** again.

Here's a close-up of the result:

Smoothly Blending in a Sky

You can use Photoshop to blend a nicer sky into a landscape photo. However, if you do it wrong, it'll look unnatural.

The most natural way to blend a sky is to use a gradient in a mask. Consider these two images. The top picture has a nice sky, but I want to use the foreground from the picture with the bird.

First, I'll set the opacity of the sky layer to 50% so that I can see through it. Then, I'll move it to align the horizons of the two photos.

Next, I'll add a layer mask to the sky.

Next, I'll press **D** to set the foreground and background color to white and black. I'll select the Gradient tool (**G**) and draw a short line from top to bottom just above the horizon. The layers should now resemble this:

A short line will create a very harsh transition, while a longer line will create a smoother transition. Experiment with the length and placement of the gradient.

Finally, I'll adjust the opacity to blend the layers naturally. If you blend smoothly enough, it might look good at 100% opacity. Here's the final result, which looks very natural.

Many photographers make the mistake of trying to select around trees and other detailed objects at the horizon, rather than just drawing a big gradient. The result is a very unnatural edge that often has ugly fringing, especially around leaves.

Masking Example: Blending Two Versions of One Photo

Here's a picture, straight out of the camera:

It's washed out. The car looks much less vibrant than it did to my eyes, and the sky looks almost solid white even though there is detail in the clouds.

You can solve this using Photoshop alone, but a quicker way is to use Lightroom and Photoshop together. In Lightroom, make two virtual copies of the picture. Then, edit one so that the car looks great, and edit the second so that the sky looks great. Here are the two examples:

Now, I'll select the two virtual copies in Lightroom, right-click them, and select **Edit In | Open As Layers in Photoshop.**

If you don't want to use Lightroom, you could just open the image in Photoshop, duplicate the layer, and adjust each layer individually for the car and the sky.

In Photoshop, I'll follow these steps:

1. Place the photo edited for the sky on top.
2. Add a layer mask to the top photo. Your layers should look like this:

3. Set the foreground and background colors to their default by pressing **D**.
4. Select the Gradient tool by pressing **G**.
5. Select the layer mask.
6. Drag a gradient by clicking in the middle of the sky and dragging to the tops of the heads of the people in the car. The picture should resemble the following:

Experiment with the gradient in the layer mask. Drawing a short gradient will show more of the vibrantly edited sky, but it will look less natural. Drawing the gradient lower will also show more of the sky, but it might change the coloration of the people.

Saving Transparent Images

Occasionally, photographers need to create transparent images. Transparent images can be dropped on top of any background, and the background will show through. Transparent images are a common element in both video editing and web design.

Let's separate the original convertible image from the background:

1. Use Select and Mask to kselect the car and save it to a new layer with layer mask, as described in Chapter 4.
2. Hide the background layer.
3. Paint white or black in the layer mask to refine the transparent convertible picture.

Here's what the image now looks like:

By default, Photoshop shows the checkerboard pattern to indicate transparency.

Before I save it, I'll crop it tight around the car to reduce the file size. It won't change the appearance of the image if you leave transparent pixels around the car, but a smaller file size will improve web page load times.

If I were to save this as a JPG file, Photoshop would automatically fill in a white background. To use transparency, I'll need to save it as a PNG file. Using the Export As dialog (**File | Export | Export As**), I'll set the file format to PNG and select the **Transparency** checkbox, as shown next.

The transparent PNG is ready for your designer. Note that transparent images won't look natural if they have a drastically different background behind them. In the case of this car, the background was very bright, so the car will look natural against a white background. It would look less natural against a black background, as the next examples demonstrate.

7 REMOVING NOISE

Watch training videos at:
SDP.io/PSEDL

Noise is that speckled digital grain you see in pictures. Every image has some noise, but it's most obvious at high ISOs, in shadows, in smooth out-of-focus areas, and in areas of solid reds and blues (like a blue sky). The next example shows a high-noise image taken at ISO 12,800.

There are many different ways to reduce noise in your photos. Ideally, you'd start the process of getting a clean image while you're taking the picture by using these techniques:

- Expose the picture so that a small part of the picture is pure, bright white.
- Use raw, rather than JPG.
- Use the lowest native ISO that will still give you a usable shutter speed.
- Bracket your exposure and combine the multiple pictures using Lightroom or Photoshop's HDR tools.

For more information about HDR, refer to Chapter 14, "HDR & Bracketing."

However, all photographers who work in low light will have pictures that can benefit from selective noise reduction. This chapter will show you how to reduce the appearance of noise in your photos without destroying the detail.

Image Averaging

If you can't bracket your exposure, another option is to capture multiple photos with the same exposure and then combine them using image averaging.

Functionally, this is exactly like shooting at a lower ISO. If you combine two pictures, your image would have about the same noise as if you had shot at half the ISO. If you combine four pictures, the noise will be equivalent to shooting at one-quarter the ISO, and so on.

For example, if you combine four pictures taken at ISO 800, the resulting image will have the same noise as a shot taken at ISO 200, because 800 / 4 = 200.

Follow these steps to reduce noise with image averaging:

1. Open the images as layers in Photoshop.
2. Select all the layers. (Click the first layer, Shift-click the last layer.)Select **Edit | Auto-Align Layers**. You can skip this step if you used a tripod when capturing your photos.
3. Select **Layer | Smart Objects | Convert to Smart Object**.
4. Select **Layer| Smart Objects| Stack Mode | Mean**.

For detailed instructions, including shooting techniques, watch the video at *sdp.io/average*.

Basic Software Noise Reduction

Though this isn't a Lightroom book, I typically use Lightroom to apply some noise reduction to all images. I typically use **Luminance** and **Color** values of 25. If the noise is more distracting than the lack of detail, I'll increase it. If I want more detail and I don't mind more noise, I'll decrease it.

This is the compromise of noise reduction: you get less noise, but you also lose detail.

Lightroom applies noise reduction to your entire photo, which isn't particularly intelligent. For example, in the photo of the bird at the beginning of this chapter, it's very important to me that the eye remains sharp. I wouldn't want to give up much detail in the bird itself.

However, in that same photo, the background is out of focus and completely smooth. There's no detail in the background at all, so I'd lose nothing if I used the maximum value for noise reduction.

Consider these two close-ups of a bird taken in low light at ISO 6400. The first shows noise reduction at 0, while the second shows noise reduction at 100.

With luminance noise, adjacent pixels have different levels of brightness, even though they should have the same brightness.

To show luminance noise, here's an image with color noise reduction at the maximum value but luminance noise reduction at 0. As you can see, the background is a single color; the noise speckles are changes in brightness only.

When you raise the luminance noise reduction too high, details disappear, as shown in the first picture in this section. When you raise the color noise reduction too high, colors disappear. If you look at the second picture, the yellow patch of feathers at the base of the bill is almost completely washed out.

The side effects of raising color noise reduction too high are typically less than the side effects of raising luminance noise reduction too high. Therefore, you can usually use a high color noise reduction value than luminance noise reduction value.

Reducing Noise in Out-of-Focus Areas

You might not notice noise in detailed parts of the picture, but it will be very obvious in areas that are supposed to be smooth, like a blurred background.

Use the Focus Area tool (**Select | Focus Area**) to reduce noise in just the defocused parts of a picture, as shown next. Typically, I leave **Image Noise Level** set to **Auto** and I adjust the **In-Focus Range** area as needed. If the tool fails to select part of the subject, I'll hold down the **Alt/Opt** key and drag the tool over the area to add it to the selection.

When the subject is well masked, I'll click **Select and Mask** to invert and further refine the selection before outputting it as a new layer with a layer mask. For detailed information, refer to Chapter 4, "Selections."

At this point, I now have a layer with just the background masked:

With that new layer selected, I can use the Noise Reduction tool (**Filter | Noise | Reduce Noise**) aggressively, setting the **Strength** value to **10** and **Preserve Details** and **Sharpen Details** set to **0**. Often, applying the filter once isn't enough. You can easily reapply the last filter you used by pressing **Ctrl+F/Cmd+F** or selecting **Filter | Reduce Noise** (which will now be at the top of the menu).

After applying heavy noise reduction to the new layer, and no noise reduction to the subject, the background is much smoother, but the subject hasn't lost any detail.

Reducing Noise in Shadows

The brightest highlights of a picture receive about eight times more light than the darkest shadows. That means that the shadows will have about eight times more noise.

Using this knowledge, you could apply heavy noise reduction to only the shadows while leaving the nicely exposed highlights completely untouched. Here's how you can do that in Photoshop:

1. Duplicate the background layer. Optionally, rename the new layer to Noise Reduction.
2. Use the Reduce Noise tool (**Filter | Noise| Reduce Noise**) to eliminate the shadow noise in the new layer. You might need to apply the filter repeatedly. Don't worry about losing details in highlights; apply as much noise reduction as you need to make the shadows smooth.
3. Duplicate the background layer again (**Ctrl+J**). Optionally, rename this new layer to **B&W**.
4. With the new B&W layer selected, choose **Image | Adjustments | Hue/Saturation**. Set the **Saturation** to -100 to make the layer black and white. Your layers should now resemble these:

5. Select the B&W layer, and copy the entire layer to the clipboard. You can quickly do this by pressing **Ctrl+A/Cmd+A** and then **Ctrl+C/Cmd+C**.

6. Select the Noise Reduction layer and create a blank layer mask. Alt-click or Opt-click the new layer mask to display it. You should see only white.

7. Paste the B&W layer you copied onto the layer mask by pressing **Ctrl+V/Cmd+V**. Your layers should now resemble the following:

8. Invert the mask by pressing **Ctrl+I/Cmd+I** or selecting **Image | Adjustments | Invert**. The brightest parts of your image will now be black in the mask, preventing the noise reduction from applying to the highlights.

9. Alt-click or Opt-click the layer mask to again display your photo.

10. Hide the B&W layer.

11. If you need more noise reduction, apply the Reduce Noise filter again to the Noise Reduction layer.

Your image should now resemble the following. The noise reduction is applied 100% to the black parts of the pictures and 0% to the white parts of the picture, with seamlessly smooth transitions between highlights and shadows.

Other Tools to Reduce Noise

Like everything you might want to do, Photoshop provides several different tools. In fact, Adobe no longer officially recommends the Reduce Noise tool because they say it's now outdated. However, I still recommend using it because I did side-by-side tests with the older Reduce Noise tool and the new methods, and for the types of selective noise reduction I'm recommending in this chapter, the Reduce Noise tool looked the most natural.

Here are two other noise reduction tools that you might want to substitute for the Reduce Noise tool:

- **Camera Raw (Filter | Camera Raw Filter)**. On the Details tab (the third tab), raise the **Luminance** and **Color** values.

- **Smart Sharpen (Filter | Sharpen | Smart Sharpen)**. Yes, Adobe hid a noise reduction tool inside of Smart Sharpen. To use this tool to reduce noise, set **Amount** to **0** and **Reduce Noise** to a fairly high value, such as **100**.

Like Reduce Noise, you might need to apply either of these tools multiple times to achieve the desired amount of noise reduction.

Unfortunately, I find that both these tools leave strange artifacts, which is why I recommend Reduce Noise instead. However, if you get better results with the new tools, you should use them.

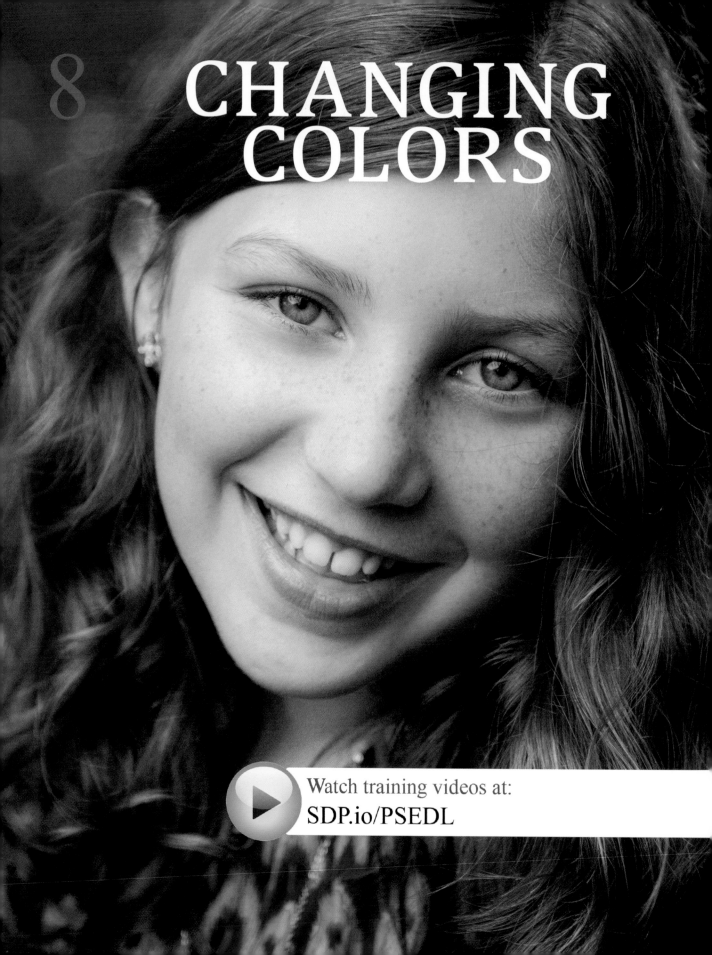

8 CHANGING COLORS

Watch training videos at:
SDP.io/PSEDL

Portrait photographers often need to change colors in a photo. If a corporate client shows up for a headshot and decides that his blue tie should be a powerful red, you can fix it in post faster than he could tie it. If you photograph a family and everyone is wearing blue, but the shades of blue are different enough that they don't quite match, you can shift the blues so everyone's outfit is consistent.

Like everything in Photoshop, there are many ways to change color. I'm going to suggest using adjustment layers because I've found them to be easy, quick, and maintainable.

Fixing White Balance

Pictures taken under artificial light will often have orange, blue, or green color casts. For example, this picture of my dog, Cowboy, is unnaturally orange because it was taken indoors, at night, under incandescent lights.

Adobe Lightroom is the easiest way to fix this—simply set the White Balance to Auto, or manually change the **Temp** and **Tint** sliders until the colors look natural.

If you prefer to make the change in Photoshop, use **Filter | Camera Raw Filter**. Then, select **Auto** for **White Balance**, as shown next. If the color still doesn't look right, manually adjust the **Temperature** and **Tint** sliders as desired.

A more precise way of adjusting the color is to use the White Balance Tool located in the top menu bar in the Camera Raw Filter, as shown next. After selecting the tool, click a part of the picture that is grey or white. Photoshop will adjust all the colors in the picture so that the area you clicked is white. For example, clicking Cowboy's grey mouth resulted in perfect white balance for the entire picture.

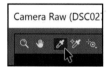

Selectively Changing White Balance

If a photo is lit by multiple light sources, you might not be able to fix the color with a single adjustment. This is common in urban night photography, which has a mix of different types of light bulbs, as well as natural light. Consider this photo of New York City (taken before the construction of the Freedom Tower):

That photo doesn't necessarily need any color changes, because it looks fairly natural. However, the lights from the city are very orange. If I manually set the white balance from different parts of the picture that should be white, I get widely varying results:

My preferred way to quickly resolve this is to simply drop the saturation of the oranges and yellows by using the Hue/Saturation adjustment layer, as shown next. Many photographers will prefer the intense oranges of the original shot, but the adjusted shot looks more like what the human eye sees.

How to Change the Color of a Dress

Consider this lovely family. Several of the group are wearing a similar shade of blue, but the shade of the teal dress is just different enough that it clashes.

There's nothing else in the photo that's the same shade as the dress, so we can easily fix this by adding a Hue/Saturation adjustment layer (as described in Chapter 2).

By default, the Hue/Saturation color range is set to Master, which will change all the colors in the photo. Her dress is closest to Cyan, so selecting that color from the list is a good start.

If the preset colors don't exactly match the color range you want to change, manually adjust the sliders at the bottom of the panel.

Now, you can adjust the Hue slider to change the color and visibly match it to the other blues in the scene.

Finally, adjusting the Saturation and Lightness can more exactly match the colors.

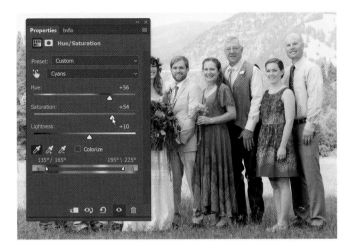

If this technique changes more of the scene than you want, apply a mask to limit the impacted parts of the picture, as shown next. For more information about masking, refer to Chapter 6.

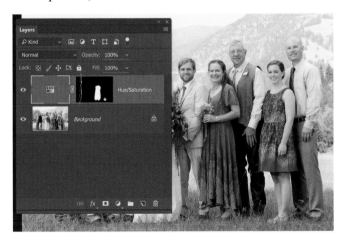

How to Add Color with Colorize

If you need to add color to a white, grey, or black subject, use the Colorize setting. Let's use it to make the bride's white dress match the other blue dresses.

First, roughly select the area to be colorized. For this example, I used Select and Mask to select the bride's dress using Quick Selection and Edge Detection, and then output the selection to a new layer with a layer mask. For more information about making selections, refer to Chapter 4.

Now, add another Hue/Saturation adjustment layer. Because the dress is selected, Photoshop will automatically create a layer mask for us.

In the Hue/Saturation Properties dialog make sure the Colorize checkbox is selected, then adjust the Lightness slider so that the bright white dress is completely colorized. Then, adjust the Hue and Saturation sliders so that the color matches the other dresses. The next figure shows the colorized results.

If you notice that the masking isn't perfect, right-click the adjustment layer mask and then use **Select | Select and Mask**. You might need to use the **Shift Edge** and **Feather** settings to more perfectly blend the colorizing adjustment layer. You'll almost always need to use some feathering to make the selection look natural.

As with everything in Photoshop, the more severe your changes, the less likely they are to look natural. In this case, changing a bright white dress into a dark blue dress is almost impossible to do perfectly, for a few reasons:

- If your masking is at all imperfect, those mistakes will be more obvious the more severe the change is.

- All objects reflect light onto their surroundings. A bright white dress acts like a huge reflector, adding fill light into shadows. As a result, the shadows near a bright dress will look very different from the shadows near a dark dress, and fixing those shadows will be very time consuming.

- Objects reflect color onto their surroundings. A blue dress would add a blue tint to everything around it, including the clothing of people nearby and the subject's skin. Most people don't consciously notice this natural occurrence, but when that reflected tint isn't there, the image seems a bit off.

Therefore, whether you can create satisfactory results using Colorize depends on the severity of your changes, your patience when creating the mask, and the keen eye of the observer.

How to Change Brown Hair to Blonde

You can change brown hair to blonde with the Levels adjustment layer and a layer mask, as this next before-and-after demonstrates.

After selecting the hair, add a Levels adjustment layer. Lower the white point to 180-200, and lower the mid-tones to around 1.5, as shown next. Adjust the levels to get the desired effect.

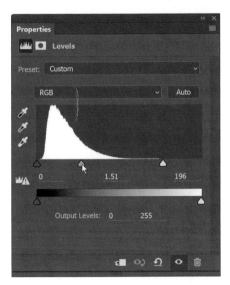

How to Add Color to Hair

If a model wants to see how he or she would look with red hair, or any other color hair, you can try it out in Photoshop before the model heads to the salon:

1. Select the model's hair.
2. Add the Hue/Saturation adjustment layer.
3. On the adjustment layer properties, select the Colorize checkbox and then adjust the Hue and Saturation as desired, shown next.

How to Change Hair to Grey

Making hair grey requires two adjustment layers, each carefully masked to select only the model's hair:

- A Hue/Saturation adjustment layer that reduces saturation to around -85.
- A Levels layer that lowers the whites to increase the brightness of the hair. For example, you might drop the white levels from 255 to somewhere between 150 and 225.

For example, this next image shows how my daughter would look if she had inherited my premature greying:

9

FIXING PERSPECTIVE
FOR REAL ESTATE
AND ARCHITECTURE

Watch training videos at:
SDP.io/PSEDL

If you're shooting at fairly close range with a wide-angle lens, perspective will cause vertical lines to appear angled in towards the center of an image. You don't necessarily have to fix all perspective distortion; our eyes see it, too, so it appears natural.

However, if you're shooting a building with a super-wide angle lens, the perspective distortion can often be disturbing. Real estate photography is often improved by fixing perspective distortion.

Fixing Perspective Distortion

Consider this real-estate photo of a kitchen. It's not bad, but the bottoms of the cabinets are tilted in towards the center of the image. That's a natural part of shooting with a wide-angle lens.

As a photographer, it's often better to fix problems like this in-camera. However, fixing this perspective distortion in-camera would require using an expensive tilt-shift lens and a tripod. Additionally, because tilt-shift lenses are available in limited focal lengths, the photographer doesn't have the flexibility to control the angle of view.

To fix the perspective in Photoshop, follow these steps:

1. Select the entire image by pressing **Ctrl+A/Cmd+A**.
2. Select **Edit | Transform | Perspective** to open the perspective transform tool.
3. Press **Ctrl+'/Cmd+'** (hold the **Ctrl** or **Cmd** key and press the apostrophe key) to show a grid that will help you align the image.
4. Now, grab a bottom corner of the picture and drag it outwards to fix the perspective correction.
5. Grab the handle at the bottom center of the image to drag it left or right, as needed.
6. Finally, press **Enter**.

You'll notice that you crop the outside edges of the photo; you can make up for that by shooting wider than needed when you're taking the photo.

Another side effect of perspective correction is that the image might seem vertically stretched or squished. To fix that, select **Edit | Transform | Scale** and drag the top or bottom points up or down until the subject looks right. Again, this will result in some cropping, so you'll benefit from using a wider angle lens than necessary at the time of shooting.

Comparing the before and after shows the differences. Most people wouldn't notice, and the improvements probably won't sell the house any sooner. However, the corrected image is subtly nicer and more professional.

Straightening Off-Center Images

If you photograph a building off-center, the far side of the building will appear smaller. The more wide-angle your lens, the more disturbing this effect will be. This real-estate shot demonstrates that:

For that shot, I would have preferred to stand directly in front of the center of the house. However, the landscaping was preventing me from getting that angle. I had no choice but to stand off-center.

You can, however, improve that shot in Photoshop. These steps will correct that off-center perspective distortion:

1. Select the entire image by pressing **Ctrl+A/Cmd+A**.

2. Select **Edit | Free Transform** to open the perspective transform tool.

3. Press **Ctrl+'/Cmd+'** (hold the **Ctrl** or **Cmd** key and press the apostrophe key) to show a grid that will help you align the image.

4. Now, grab a corner on the right side of the picture and drag it outwards while holding the **Ctrl/Cmd** key.

5. While still holding the **Ctrl/Cmd** key, grab the handle at the right center of the image to drag it up or down, as needed, to re-center the house. Fix the perspective correction by aligning the lines of the roof to the gridlines. As shown here, the house will look squished:

6. Press **Enter**.

7. Select **Edit | Transform | Scale** and drag the center right handle outwards to fix the squished perspective. Press **Enter**.

Comparing the before and after photos, you can see that correcting the perspective creates a more natural-looking image. Of course, the fixed image is actually less natural, but it's more like we'd see the house in our mind's eye.

Using Lightroom to Fix Perspective

Though this book is about Photoshop, I need to mention that you can make many of these changes much more quickly in Adobe Lightroom by using the Transform panel in the Develop module, shown next. Click **Full** or **Auto** to make Lightroom take a guess at how to correct the image

10 CHANGING SHAPE WITH LIQUIFY

 Watch training videos at:
SDP.io/PSEDL

Liquify can change the shape of subjects in a photo; it's one of Photoshop's most powerful features. To open Liquify, select **Filter | Liquify**.

In this chapter, I'll show you the basics of using Liquify. However, Liquify is more art than science. It simply takes practice to get great results.

Standard Liquify Tools

Here's a quick overview of how to use the standard Liquify tools.

Forward Warp Tool

You'll spend most of your time in Liquify using the Forward Warp tool. Simply click and drag to change the shape of any subject.

For instance, a few pushes with the Forward Warp Tool gave this house finch a much prouder chest, without any editing artifacts:

Selecting the correct brush size (using **[** and **]**) and pressure is key to getting great results with this tool. In general, a larger brush with lower pressure will create more natural-looking results. Your brush should almost always be bigger than the area you want to change, so that the changes blend smoothly into the rest of the image.

Reconstruct Tool

The Reconstruct tool removes any Liquify edits. You'll use it often because once you start making changes in Liquify, it's hard to avoid going too far with them.

Basically, when you reach that point where you know you've gone too far with Liquify, select the Reconstruct tool to reduce the severity of your changes. The default **Pressure** of **100** completely removes all changes; I typically keep the **Pressure** set at **5-10**.

Smooth Tool

If your edges look rough, use the Smooth tool to even them out.

For example, this next before-and-after example shows how the smooth tool evens out rough edges from a bad forward warp.

By default, the pressure on the **Smooth** tool is set to 100. That's typically far more than is required; a pressure of 20-30 is more useful.

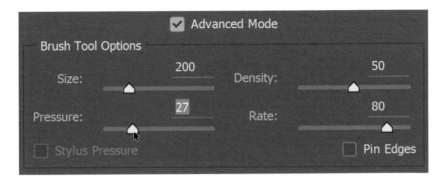

Twirl Clockwise Tool

The Twirl Clockwise tool rotates subjects.

For example, I could use it to slightly change the angle of this chipmunk's eye. To rotate counter-clockwise, hold down the **Alt** or **Opt** key. The next examples show the original image, the image with the eye rotated clockwise, and the image with the eye rotated counter-clockwise.

As with other Liquify tools, you'll get the most natural results when the brush is far larger than the area you actually need to rotate.

Twirling is useful in many scenarios. For example, you might use it to slightly change the angle of the head tilt for a portrait model, or to make a hand pose at a more attractive angle.

Pucker Tool

The Pucker tool squeezes in part of the picture, making a feature smaller.

For example, I used the Pucker tool to give this chipmunk small, beady eyes. Here's the before and after.

As with the other tools, use a brush much larger than the feature that you want to modify.

Bloat Tool

The Bloat tool is the opposite of the Pucker tool. Use the Bloat tool to make a feature larger.

For this next example, I used the Bloat tool to give the chipmunk larger eyes.

Push Left Tool

The Push Left tool can create a similar effect to the Forward Warp tool; it simply pushes part of the picture. In most cases, the Forward Warp tool is more useful.

The Push Left tool pushes the image to the left, relative to the direction you move your mouse. If you imagine that the cursor is a car, and you're driving it around the screen with your mouse, Liquify will be pushing everything to the left relative to the car.

It's a really strange way to interact with your computer; I can't think of a single other application that uses the relative direction of the cursor. Here's how it behaves:

- If you drag your cursor up, Push Left will push everything to the left.
- If you drag your cursor down, Push Left will push everything to the right.
- If you drag your cursor left, Push Left will push everything to the down.
- If you drag your cursor right, Push Left will push everything to the up.

It's weird, but it is useful for expanding or shrinking a subject's full body. For example, on the house finch we saw earlier, I used a brush size of 1600, a brush pressure of 17, and only two mouse strokes (one up, one down) to make the house finch much fluffier. Here's the before and after:

Freeze Mask & Thaw Mask Tools

Use the Freeze Mask tool to prevent Liquify from changing part of a picture. Use the Thaw Mask tool to unfreeze part of a picture.

When you change the shape of a subject, Liquify stretches the surrounding area. That stretching can make it obvious that changes were made. Therefore, the Freeze and Thaw tools are really important for preventing the background from changing when you modify the shape of a foreground subject.

In this next photo, the model's left shoulder seems to be sticking out unnaturally. I would normally fix that with the Forward Warp tool. However, the geometric lines in the background become unnaturally warped, as this next before and after example demonstrates.

When I froze the straight window panes, Liquify left them untouched and stretched only the surrounding area. The next two examples show the same area stretched with the freeze mask shown and hidden.

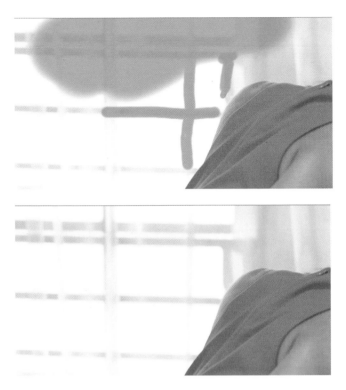

Use the **Show Mask** checkbox to display or hide the red mask. You can also change the color of the mask, which is useful if your image is red.

With this chipmunk, any unnatural curves added to the hard, straight cuts in the wood would make the use of Liquify obvious. To prevent Liquify from stretching that part, no matter which change I apply, I'll paint it with the Freeze tool, as shown in red below. If I overpaint part of it, I'll use the Thaw tool to erase the Freeze tool.

Face Tool

Use the Face tool to quickly make subtle edits to people's faces and expressions.

The Face tool attempts to identify every face in the photo. You can then select a face by clicking the face, and make adjustments to the eyes, nose, mouth, and face shape. It works so well that it's a bit creepy.

In the next example, this couple didn't need any editing. However, I edited their faces just to demonstrate what the types of subtle changes the Face tool can do in just a few minutes. Philosophically, I don't believe in changing the shape of people's faces for portraiture. However, the Face tool is still useful for correcting awkward expression caused by squinting, discomfort, or bad timing.

The Face tool is quick and easy, and can drastically reduce the time a portrait photographer spends editing photos. Some portrait scenarios will still require using the Liquify tool manually. For detailed information, read "Manually Changing Faces" at the end of this chapter.

Hand Tool

If you've zoomed into the picture so that it can't all fit on the screen at once, use the Hand tool to drag the visible part of the image around. It works exactly like Photoshop's standard hand tool.

Just like in Photoshop, I find it easier to hold the Space bar to quickly and temporarily select the Hand tool rather than actually click it.

Zoom Tool

Select the Zoom tool and then click to zoom into the photo. Alt-click or Opt-click to zoom back out.

I use the scroll wheel of the mouse to quickly zoom in and out of a photo, or you can press **Ctrl++** and **Ctrl+-** or **Cmd++** and **Cmd+-**.

Reconstruct Options

You can always click **Load Last Mesh** to undo all your changes in Liquify. If you feel you've slightly overdone all the changes, you can use the Reconstruct button under **Brush Reconstruct Options** to scale back all your changes.

I will always use the Reconstruct tool to see if scaling back my changes a bit looks better. Most editors have the tendency to overdo Liquify, so dropping the amount to 60-90 before saving your image reduces the chances that your edits will be obvious to others.

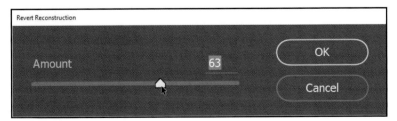

The Reconstruct tool is also a great way to show other people your before and after. Drag the slider up and down to animate your changes.

Showing the Mesh

Located under View Options, select the **Show Mesh** checkbox (shown next) to display a grid where Liquify has stretched the image.

Viewing the mesh is a good way to verify that you haven't stretched part of the picture in such a way that it will look obvious.

Using Advanced Mode also gives you access to the **Load Mesh** and **Save Mesh** buttons. These buttons save different Liquify configurations. However, I personally prefer to use the Photoshop History, Snapshot, and Layers to save different versions of Liquify edits.

Changing Clothes and Bodies

Photoshop has tools that can fix wrinkles in clothes or change the shape of people's bodies. As a portrait photographer, I was often asked by clients to make them appear thinner. As a commercial photographer, I was often required to make every model look commercially perfect.

Some uses of these tools are understandably controversial. Certainly, publishing unrealistically edited images of people's bodies has contributed to emotional conditions such as body dysmorphia, which has negatively impacted lives. For many photo editors, these skills are a requirement; thus, it's a requirement for me to teach them. However, I ask that you be aware of how unrealistic images can impact the people viewing the picture and that you use your skills responsibly.

Here's a studio shot of a model. Because of the way she's holding her arms and the fact that the dress fits her loosely, the dress is hanging away from her body and hiding her waist.

Ideally, this would have been fixed at the time of shooting by pinning the dress behind her back. However, it can also be fixed with Photoshop by following these steps:

1. Select **Filter | Liquify**.

2. Select the **Forward Warp Tool** in the upper-left corner.

3. Select a nice, big brush size with a moderate pressure. For this sample photo, I chose a brush size of **600** and a pressure of **20**.

4. Drag with the brush from the outside in to gently change the shape of the dress, as shown next.

Comparing the before and after photos, you can see that the change is subtle and that we weren't changing the shape of her body, but simply correcting the photographer's oversight of the poorly fitted dress.

Changing Faces

You can also change the shape of facial features using the Liquify tool in Photoshop. For this next example, we'll start with a photo of a model recreating a scene from the movie *Scarface*. The model resembles the lead actor from that movie, but a few tweaks to his facial structure can make him look even more similar.

1. Select **Filter | Liquify**.

2. First, let's adjust the model's jawline so it more closely resembles Tony Montana's. The model's jaw is more sharp and square, but pushing on the corners of his chin and jaw with the Forward Warp Tool can fix that. This time, a smaller brush is appropriate. Short, subtle changes work best.

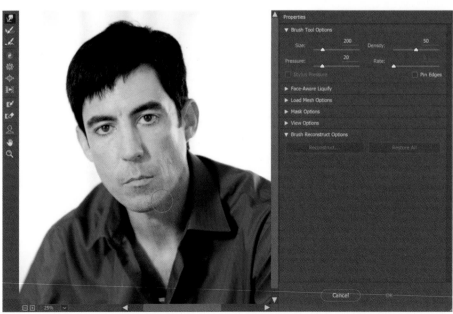

3. The movie character had a wider mouth than our model has. Again, use the forward warp tool to push the corners of the model's mouth outward. If you pushed the corners of the mouth up or down, you could use this technique to make the model's expression more or less happy.

4. Next, we'll add a bit of droop to the outside corners of the model's eyes. For the first pass, a large brush of 300-400 is useful because it will change the skin around the eye and even the eyebrow, which is more natural. To refine the shape further, use a smaller brush of around 150.

5. Finally, I'll use a huge brush of around 700 to slightly lower the model's hairline.

Comparing the before and after, the individual changes are subtle, but cumulatively they make the model look like a different person.

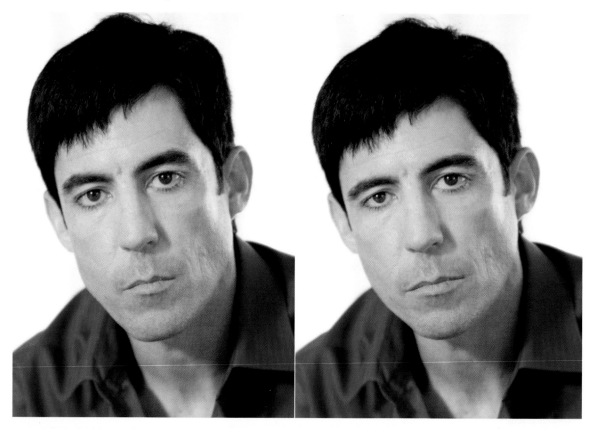

Changes to human faces need to be very subtle because we humans notice even the slightest distortion to a person's face. Especially when you're new to using Liquify, it will be tempting to make major changes. Err on the side of smaller changes.

As a portrait photographer, you'll often have clients ask you to make them thinner or to make their noses smaller. Even though it's the client's wish, be careful about the changes you make. While the client might be happy with it, if the change is obvious, other people will notice.

If a client does want part of his or her body to appear different, it's preferable to compensate for that with posing and lighting. As a last resort, use Photoshop sparingly.

11 REPLACING BACKGROUNDS

Watch training videos at:
SDP.io/PSEDL

This chapter provides tips and techniques for replacing the background or sky in a photo, and is useful to landscape, wildlife, and portrait photographers. Before reading this chapter, you should have a good understanding of making selections, layers, and masks (covered in chapters 4-6).

Matching the Light, Focal Length

The more similar your new background is to the original photo's, the more realistic it will look. For best results, choose a background with these characteristics (as compared to the original photo):

- **Similar focal length**. If you try to add a background taken at 24mm into a picture taken at 100mm, it's going to look unnatural. For best results, use a background taken with a similar focal length.

- **Similar brightness**. Because background light interacts with foreground subjects, replacing a dark background will a light background will look unnatural, and vice-versa.

- **Similar light direction.** If you have a front-light portrait and your replacement background is back-lit or side-lit, the multiple light sources will look unnatural.

- **Similar light hardness**. If the foreground is lit with hard light, the background should be, too. If a picture was taken on an overcast, the replacement background should also be taken on an overcast day.

- **Similar angle.** Don't try to replace the background of a photo taken while pointed at the horizon with a photo taken while pointed directly at the sky.

What happens if you don't follow these guidelines? First, consider this picture of Porto, Portugal, taken at 70mm with a full-frame camera:

The sky is rather boring. Let's try to replace it with a more dramatic sky, taken as a rainstorm was rolling in:

Even at a glance, it doesn't look realistic. The ocean on the left of the photo is brightly lit, which doesn't match the dark sky. As we zoom into the details, inconsistencies become even more obvious:

Notice the haloing around the posts. Some of this could be addressed by very carefully masking the posts, but it will never completely go away, because backlighting actually makes textured objects glow. Replacing a light background with a dark background will always look weird.

Take a closer look at the lighthouse:

I carefully masked around the cabling on the lighthouse, but the backlit cabling is slightly glowing because of the original black background. That happens at a sub-pixel level; it's impossible to make it look realistic with simple masking.

Here's the best I could do to make the background look natural by using Select and Mask on the layer mask. As you can see, I set Shift Edge to -77% to squeeze the selection further into the posts, causing the background to cover the edge of the posts. I also set Feather to 1.9 pixels to blend the background into the edges.

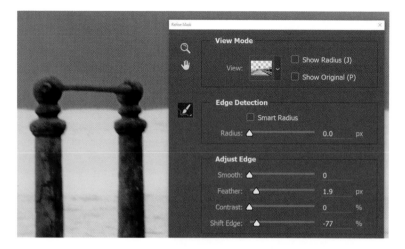

That looks fairly natural, though shifting the edge has slightly changed the shape of the posts, and there's still a bit of a glow. However, look what that Shift Edge function did to the lighthouse:

Shift Edge worked well on the larger subjects, but completely destroyed the finer parts of the mask. You can work around this by applying Shift Edge differently to different parts of the image, but that significantly increases your post-processing time, and some parts of the image will never look right.

Here, I used a replacement background that more closely resembled the original: it's bright and taken at about the same focal length. While it's not perfect, most people wouldn't notice that it had been replaced. Yet, it's different enough to change the look of the picture.

Looking again at those posts, the replacement background looks very natural:

What Gives Away a Fake Background

Unless a background is completely black, the background is, itself, a source of light. A bright background causes the edges of subjects to slightly glow. Backgrounds with color cast that color onto foreground subjects.

This is fairly obvious, but few people think about it. Anything you take a picture of is sending or reflecting light towards your camera; that's why your camera can capture its image. Some of that light also strikes objects in the foreground of your picture, adding brightness or color to them.

That light interacts with the foreground subjects in subtle ways that you might not consciously notice. Unconsciously, however, a viewer will notice if the foreground has a background light source that isn't present in your final picture because you replaced it with a darker background. They'll also notice if the foreground is missing a background light source because you replaced it with a lighter background.

Differences in color matter, too. If you replace a red background with a blue background, the background itself might look realistic, but light reflected from the background onto the foreground subject might not change in the same way.

Selecting Around Leaves

If you replace a background in a landscape photo, there's a good chance the leaves of a tree will appear against the sky. Leaves have far too much detail to select individually.

Consider this portion of a sample photo, which shows leaves against a dark cloudy sky.

If you decided to replace this with a clear blue sky, you could do so convincingly by using the Color Range selection tool (**Select | Color Range)**. Select the color of the sky behind the leaves by shift-clicking parts of the sky. Then, adjust the Fuzziness and Range as desired.

The next figure shows the blended selection against a fake blue sky that I created by drawing a gradient from light blue to white. As you can see, the final product looks natural. However, the original image had blue clouds behind the leaves, so the replacement color isn't very different from the original color.

Here's a bigger challenge: replacing a bright white cloud with a darker blue sky.

Using the same Color Range selection technique leaves harsh fringing:

Again, using the Select and Mask tool to feather and shift the edge can reduce the fringing and improve the results. However, some thinner branches completely disappear. The results might be good enough for sharing on the web, however.

Making Backgrounds for Future Use

For best results, take photos of nice backgrounds specifically to use in future photo projects. For example, if you have a particularly sky, take photos of it at a variety of focal lengths (such as 16mm, 24mm, 50mm, 80mm, 100mm, and 150mm).

If a scenic landscape lacks a focal point, it might be an excellent background for use in a future portrait. Because you don't know what image you might be blending into it, take pictures of the background with a variety of different focal lengths, and shoot both vertically and horizontally.

Blurring Appropriately

Backgrounds are almost always at least slightly blurry; one problem with replacement backgrounds is that they're often sharper than they would be in the original picture, and our human brains can detect that difference.

If you're photographing a flat background, such as the sky, you don't have to worry about where you focus; just focus on the background. When you later blend the background into a different photo, use the Lens Blur tool to match the background blur of the original photo.

If you're photographing a background with some depth (for example, a beach scene) you might intentionally focus in the foreground to cast the background out-of-focus. For example, if you plan to use a background scene to replace the background in a portrait, imagine where your portrait subject would be standing, and focus on that spot. Then, take pictures at a variety of different f/ stops, such as f/2.8, f/5.6, f/8, and f/11.

When you later blend the background into a photo, you'll be able to choose the background that most closely matches the settings you were using when you took the portrait.

Green-screen/Chroma key

Many people think of chroma key when they think of replacing background. Chroma key is that green or blue screen that the weatherman stands in front of so the live news broadcast can easily show a different display.

Chroma key is useful for video because it doesn't require much computing power to replace a solid color in an image, and video requires updating 30 frames per second in real time. Chroma key often leaves green or blue reflections on the clothes or skin of the subject, however, and thus it's generally unsuitable for photography.

Instead of using chroma key, simply place your subject on a black, grey, or white background. Try to match the final background as closely as possible in brightness and color.

12 COMPOSITING

Watch training videos at:
SDP.io/PSEDL

Compositing, often abbreviated to comping, combines multiple photos together to create an image that might pass as authentic. Compositing is useful for many types of photographers:

- Portrait photographers could blend faces from two different group photos to create a single image where everyone has a nice expression.

- Fine arts photographers could create a complex vision that combines elements from multiple pictures.

- Commercial photographers could shoot a model in a studio environment and then comp him or her into different environments, such as an office or a beach.

- Wildlife photographers might combine multiple photos of animals to tell a story that they were not able to capture with a single shot because of limitations of their lens or depth-of-field.

All the concepts in the previous chapter apply equally well to compositing: combined images will look better when the light hardness, light direction, focal length, and brightness are similar.

Throughout this chapter, I'll work with the following sample vacation image and then add more elements to it to tell a story. I encourage you to work along with me, using the sample files. The techniques we'll apply are generally applicable to many types of composites.

Note, however, that compositing is more art than science. Because of that, you might learn better by watching this book's videos at *sdp.io/psedl*. I can't give you a step-by-step process to follow to make a composite look realistic, because every picture is different. The best I can do is demonstrate various techniques that I've found useful.

Adding and Masking People

The Trolley seems rather lonely, so let's add to the image my two favorite people: Chelsea and Madelyn. I have another photo of them walking, and it happens to have been taken under similar overcast lighting, which should make the images easier to blend.

First, add the image with the people as a new layer above the trolley picture. This can be accomplished by selecting both images in Adobe Lightroom, right-clicking, and selecting **Edit In | Open as Layers in Photoshop**. Alternatively, you could open the images separately in Photoshop and then copy and paste the image with the people on top of the trolley picture. For more information, refer to Chapter 5, "Layers."

Now, we need to isolate the people from their background. The easiest way is to use the Quick Selection tool followed by the Select and Mask tool. Save the selection to a new layer with a layer mask. For best results, feather the edge of the selection by several pixels. For more information, refer to Chapter 4, "Selections."

Matching Scale

Now, the composite image looks fairly terrifying, because the scale of the people doesn't match.

You can fix this with **Edit | Transform | Scale**. Drag the corners of the layer while holding the **Shift** key to keep the aspect ratio locked. Match the size of the people to the scale of the trolley. You know that the door on the trolley would be about a foot taller than either of the people but that the people are closer to the camera, meaning perspective would make them appear larger. Therefore, their height should be about the same as the trolley door. Press **Enter** to finalize the transformation when you've matched their scale.

Your image should now resemble the following, and already it looks much more realistic.

Changing Direction

The image feels uncomfortable because the people are looking off to the left. Not only is this violating the compositional rule of space, but if there were a fast-moving trolley a few feet away from them, they'd definitely be looking towards it.

We can address this by flipping the layer horizontally and then repositioning it. Select **Edit | Transform | Flip Horizontal**. Then, press **V** to select the **Move** tool, and drag the people back to the left side of the frame.

Now, the image resembles the following. When you flip an image, zoom in carefully and look for any writing, which would appear backwards in the flipped image.

Refining the Mask

There's a bit of haloing evident around the people because my masking wasn't perfect. That's OK, because it's better to make final adjustments to the mask after your subject is in the final position; the haloing will only be evident where the new background doesn't match the original background.

As discussed in Chapter 6, "Masking," you can paint black or white into the People layer mask to hide visible edges. Use the edge of a fairly small, soft brush and paint black along every visible halo.

A faster way to accomplish this is to use the **Select and Mask** tool, and then adjust **Shift Edge**, as described in Chapter 4. However, that tool often adjusts all edges universally, and typically goes too far on some parts of the image. You almost always need to manually paint some parts of the edges of a composited photo.

After manually removing the haloing, the blended image looks much more natural.

Matching Color

Now, you can adjust the saturation and white balance of the two composited images so they appear to match. Pictures might have different color if they were captured with different light sources, with different cameras, or at different times of day. For detailed information, refer to Chapter 2, "Adjustment Layers."

Placing a Bird Photo

Compositionally, adding people to the lower-left corner of the photo has made it very unbalanced. We can offset this by adding a contrasting subject to the upper-right corner.

Seagulls were extremely common in this area, so let's add a couple of bird photos as new layers and scale them to the right size.

Additionally, it might look nicer if the bird were flying in the same direction as the trolley:

1. Choose **Select | Select All** (or press **Ctrl+A/Cmd+A**).
2. Choose **Edit | Transform | Flip Horizontal**.

Removing the Background

First, we need to remove the blue sky from the background:

1. Choose **Select | Color Range**.
2. Set the **Range** to **100%** (so that it covers the entire layer).
3. Select the Invert checkbox so that we select the bird, rather than the sky. If you forget this step, you can press **Ctrl+Shift+I** or **Cmd+Shift+I** to invert the selection after making it.
4. Click the blue sky near the bird.
5. Adjust the Fuzziness so that the Color Range tool selects most of the blue, including the blue that's overlapping the bird itself.

6. Click **OK** to create the selection.

7. Create a mask based on the select by clicking the **Add layer mask** button.

8. Unfortunately, the bird still has a bit of blue around his edges, which looks unnatural.

9. We know seagulls aren't blue, so we can fix this by simply desaturating the blues. In the Layers panel, select the seagull image instead of the mask. Then, select **Image | Adjustments | Hue/ Saturation**.

10. In the **Hue/Saturation dialog**, select **Cyans** from the list, and then set the **Saturation** to **-100**. This won't get the entire color range in the sky, so drag the left and right sides of the color slider so that it covers the sky's entire range. Specifically, you'll need to drag the right side of the range further to the right, towards the deep blues. Click **OK**. This could also be done with an adjustment layer.

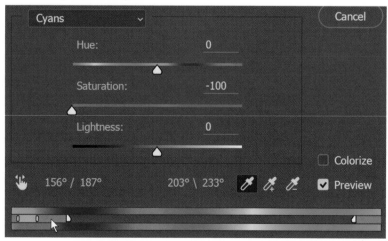

Now that we see the image more clearly, we can better resize the bird with **Edit | Transform | Scale**.

Adding Motion to the Birds

I photographed the trolley with a 1/20th shutter speed, but I photographed the seagull with a 1/250th shutter speed. While the seagull has some motion blur, it's not enough—indeed, it should have moved 12.5 times further.

While it's almost impossible to remove motion blur with Photoshop, it's really easy to add it:

1. Select **Filter** | **Blur** | **Motion Blur**.
2. Set the **Angle** to match the angle of motion.
3. Set the **Distance** to whatever looks natural.
4. Click **OK**.

Technically, you could objectively assess the motion by counting the number of pixels of movement in the original image and multiplying it by 12.5. But this is more art than science, so just adjust it however you like.

In my photo, I repeated the process for a second image of the same bird because two birds seemed to better balance the visual weight of the photo's composition.

Adding Motion to the Feet

The photo of Chelsea & Madelyn walking was taken at $1/60^{th}$, and it has some visible motion in it—particularly their rear feet, which are at the fastest part of the stride.

However, the train photo was taken at $1/20^{th}$. Therefore, for the movement of the feet to look realistic, the feet should have three times more motion. To simulate this, we can use the same filter that we used for the birds but apply it selectively to only the feet.

1. Duplicate the layer with the people. Call the new layer **Feet Moving** (or something similar).
2. Select the new layer, and then select **Filter | Blur | Motion Blur**.
3. Configure the Motion Blur filter to create realistic movement. Here, it helps to have an understanding of how people move. At the back of the stride, your foot is mostly moving upwards, so the **Angle** setting should be almost vertical. Adjust the **Distance** to look realistic.
4. Click **OK**.

Now, the entire layer has vertical motion, but that's not realistic. The rear feet are moving upward at the end of the stride, but the rest of the body is moving forward, not upward.

To show only the feet moving, use a layer mask. Hold down the Alt/Opt key while clicking Add Layer Mask to create an all-black layer mask. Then, paint white over the rear feet to show only that part of the Feet Moving layer. Now, the feet and layers should resemble this:

Adding Motion to the Bodies

The rear feet are moving the fastest, but the rest of the bodies are moving forward and need a

bit more blur to look realistic near the fast-moving trolley. However, their motion is a bit more complex.

The birds were moving side-to-side relative to the camera. Therefore, their motion would simply be blurred sideways. The rear feet were moving upwards relative to the camera. Therefore, the motion of the rear feet is simply blurred vertically.

However, the people are moving away from the camera. If they were to continue moving over a long period of time, their wouldn't simply move in one direction relative to the camera; they would get smaller as they moved towards a point of convergence on the horizon.

Fortunately, the Path Blur tool makes it simple to add three-dimensional blur. Let's try it out:

1. Select the People layer, which should be below the Feet Moving layer.

2. Select **Filter | Blur Gallery | Path Blur**.

3. In the Blur Tools panel, click and drag your mouse to set multiple points on the people and the direction they would be moving towards the horizon, as shown next. Adjust the **Speed** and **End Point Speed** settings to create a realistic amount of movement based on the speed of the subjects.

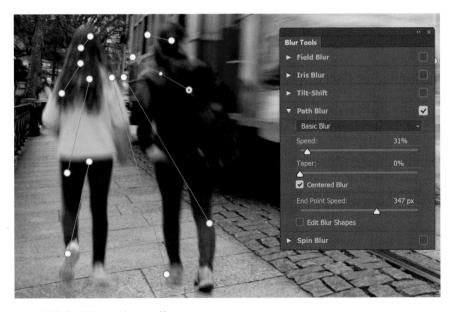

4. Click **OK** on the toolbar.

We separately added motion to the birds, the people's feet, and the people's bodies. We had to do this because the images we were comping were taken at different shutter speeds, and the slow shutter speed of the primary image was obvious because of the movement of the trolley.

We had to separately add motion to the feet because even a single subject can have motion in different directions and with different speeds. There's no formula I can provide that will tell you how to do this for every picture; you simply need to think about the mechanics of movement and how to best simulate that motion in Photoshop. Then, you need to weigh the amount of time it will take to precisely simulate that motion vs. roughly simulating the motion using fewer layers and less precision.

If there were no visible motion in your image, you wouldn't need to add any, even if the shutter speeds were different. For example, if I had used shutter speeds of 1/1000, 1/8000, and 1/500, all subjects would probably have appeared still. Not having to add motion to the images would have reduced the processing time; however, that extra motion also helped to blend the composited images together, reducing tell-tale signs of a composite at the edges of the subjects.

Adding Shadows

We have motion in the feet now, but if you look at that part of the picture, it still doesn't seem quite right. The feet seem to be superimposed over the ground, rather than actually touching the ground.

The reason is simple: they don't have shadows. Fortunately, it's easy to add shadows.

1. Select **Layer | New | Layer**.
2. Set the **Mode** to **Darken**, set the **Opacity** to about **50%,** and select the **Fill with Darken-neutral color** checkbox. Click **OK**.

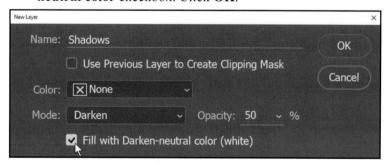

3. Move the Shadows layer below the People layer.
4. Now, we need to paint black on the new layer to add shadows. Press **D** (to select the default foreground and background colors) and then **B** (to select the Brush tool).
5. Set your brush hardness to **0%**.
6. Paint black spots underneath the people's feet where shadows would be. Soften the edges of the shadows by reducing the **Opacity** of your brush to **33%**. This picture needs very soft edges to the shadows because the original photo was taken on an overcast day. If the photo had been taken on a sunny day, the shadows would have sharply defined edges. Studying light and shadow will help you create more realistic composites.

7. The shadows will look too dark now. Reduce the **Opacity** of the Shadows layer until the shadows look more realistic. I settled on an **Opacity** of **18%**.

8. Now, fine-tune the shadows. If your shadows are too large, press **X** to switch your brush to white, and then paint over the edges of your black shadow using a brush with a low **Opacity**.

This is what I settled on for my shadow treatment. The shadows are subtle because overcast days have almost no shadows. In fact, most people would not consciously notice the shadows, but the image definitely looks more natural with the layer enabled.

Hiding Your Artifacts

At this point, your composite might still look a bit fake to you. First, take comfort because others might not notice imperfections that seem glaring to you—you've been staring at the photo for too long, and you already know it's a comp.

There's one last trick that can hide any remaining artifacts, and that's to apply toning and other special effects to your entire image. Because these effects impact all composited parts of the image equally, they serve to further blend them together.

To finish this picture, I brought it back into Lightroom and applied a preset that I had made—the Night-Super Pop preset that's included with my Lightroom books. Then, I added some post-crop vignetting and grain to the image.

It's not perfect, but it's good enough to pass casual inspection, and it tells a more interesting story than the original image did. I shared it with my Instagram followers, who consist of 16,000 photographers, many of whom are always quite eager to point any flaws in photos, and nobody noticed that it was a comp. Mission accomplished.

The final lesson about comping is that your image will never look completely real because it's not real. If you're sharing a high-resolution image for a commercial client or making a large print, you might need to spend days creating a perfect comp. If you just want to artistically combine elements to share a vision on the web, you can probably do a passable job with less than an hour at the computer.

Matching Noise Levels

If you're blending shots taken at different ISO settings, or taken with different cameras, the different images might have visibly different noise levels, and that difference will make the composite less natural.

If part of the image is scaled bigger or smaller, this will impact how visible the noise is. Scaling one of the composite images down to a smaller size reduces the visibility of the noise, while scaling a composite image to a larger size increases the visibility of the noise.

You can address this in two ways:

- **Reduce noise in the noisier image**. As described in Chapter 7, "Removing Noise," you can reduce the noise in the noisier image until it matches the noise in the rest of the image.

- **Increase noise in the cleaner image**. Use **Filter | Noise | Add Noise** to artificially add noise to match other parts of the composite picture. Experiment with the **Amount** slider and the **Monochromatic** checkbox. Usually, selecting **Monochromatic** generates more realistic results.

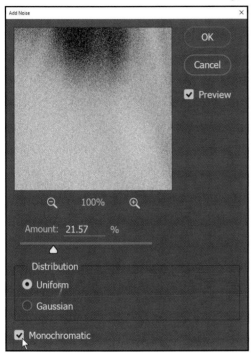

Matching Sharpness and Blur Levels

Most composites have subjects of varying degrees of sharpness, and that makes them less natural. Several factors can change how sharp a subject appears in a picture:

- **Atmospheric conditions and camera distance**. Subjects that are closer to the camera always appear sharper, especially on hazy or humid days.

- **Lens sharpness**. If you combine images taken with different lenses, one of the lenses might be sharper than the other. Even if the images are taken with the same lens, varying f/stops can dramatically change sharpness.

- **Depth-of-field**. Subjects in front of or behind the focal plane will be slightly out of focus. If you place a subject outside the focal plane of the background image, you need to add lens blur.

- **Motion blur**. As discussed earlier in this chapter, if shutter speeds vary between the different composited images, you might need to match the motion blur.

- **Camera shake**. Slight amounts of camera shake can occur in any handheld photo, regardless of the reciprocal rule.

Often, you can reduce blur (including motion blur) by using the **Filter | Sharpen | Smart Sharpen** tool. You might be able to reduce camera shake by using **Filter | Sharpen | Shake Reduction**.

At some point, sharpening and reducing shake will begin to look unnatural; don't push those tools too far. Instead, add matching blur to the other components of the composite. Use **Filter** | **Blur** | **Motion Blur** for simple motion blur or camera shake, or **Filter** | **Blur Gallery** | **Path Blur** for more complex motion blur, as discussed earlier in this chapter.

If you need to make a wheel spin (such as when making a still car appear to be moving), use the **Filter** | **Blur Gallery** | **Spin Blur** tool.

Changing Light Quality (Hard and Soft Light)

If part of your composite was taken in hard light and part in soft light, the resulting composite will look unnatural. While it's best to use pictures taken in matching lighting conditions, you can partially address varying light by using the Hard Light and Soft Light layer blending options. For detailed information, refer to Chapter 5, "Layers."

13PANORAMAS

Watch training videos at:
SDP.io/PSEDL

Panoramas stitch multiple photos of a single scene together to create a wider-angle, higher-megapixel image. Panoramas are an extremely useful photographic technique, allowing an inexpensive camera to produce better results than even the most expensive cameras.

This chapter will provide a brief overview of how to capture a panorama and then how to process it by using Photoshop. Because Photoshop is often not the ideal panorama processing software, I'll also recommend other software that you might prefer.

Capturing a Panorama

This book is focused on post-processing, rather than capturing images. For detailed information about capturing panoramas, refer to Chapter 2 of *Stunning Digital Photography*.

However, if you're already familiar with creating panoramas, here are some tips that will make post-processing much easier:

- **Stop the camera and wait one second between photos**. Often, people snap photos while continuously panning their camera. This will add motion blur to the image, reducing sharpness.

- **Use continuous shutter and take multiple photos of each section.** If a single photo is blurry (for example, from camera shake or motion blur), the entire panorama is ruined. Therefore, I like to take three photos of each section of the scene. Then, I can combine only the sharpest photos for the panorama.

- **Use autoexposure. I** know **that every other educational source will tell you to use the same exposure for all photos.** That technique can work, especially if you shoot raw, rather than JPG. However, modern panorama-processing software automatically adjusts the exposure of shots to match. Autoexposure will capture more total dynamic range in the scene, adjusting the exposure down for bright skies and up for portions of the scene that are mostly shadow. Autoexposure panoramas can produce superior results, essentially creating a high dynamic range scene.

- **Use manual focus**. While software is great at blending together images with different exposures, it's terrible at blending together images with different focal planes. Manually focus about 1/3 of the way between your foreground and background.

- **Use a higher f/stop**. Depth-of-field can be shallower on a panorama than it would be if you captured the entire scene with a single wide-angle photo. For example, if you were capturing a panoramic scene equivalent to 16mm, f/2.8 might provide you sufficient depth-of-field with a proper 16mm lens. However, if you used a 50mm lens at f/2.8 to capture that same field of view, the resulting image would be equivalent to using a 16mm f/0.9 lens.

- **Shoot raw.** Especially on sunny days, panoramic scenes will have more dynamic range than any single shot. Therefore, you might need to recover shadows or highlights in the scene, and raw files do that better than JPG files.

- **Take two extra rows and columns of pictures**. Always shoot wider than you think you need to; corners will get cut off in processing, and you will almost always need to crop your final panorama.

Using Adobe Lightroom

Whenever possible, I prefer to create panoramas in Adobe Lightroom. Lightroom tends to be faster and easier. Additionally, if you're merging raw images, Lightroom will create a DNG panorama that contains all of the information from the original photos. That allows you to recover highlights and shadows with optimal image quality.

Within Lightroom, select the images that will make up your panorama by clicking the first image and Shift-clicking the last image. Then, right-click one of the images, select **Photo Merge**, and select **Panorama**.

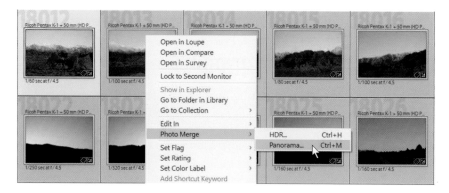

However, Lightroom tends to be more picky than Photoshop about the photos that it will process into a panorama. As a result, Lightroom often returns a vague error message. In those circumstances, I tend to open Photoshop to create the panorama directly.

Stitching a Panorama

Within Photoshop, follow these steps to create a panorama. You can retrieve the sample files from *sdp.io/psedl*. Select **File | Automate | Photomerge**. You don't need to create a new document; Photoshop will make one for you.

1. Click **Browse** to select your files. Select multiple files by **Ctrl/Opt**-clicking each file, or by clicking the first file and Shift-clicking the last file.
2. Select all four checkboxes at the bottom of the dialog.
3. Leave **Auto** selected. If that fails, try selecting **Cylindrical**.
4. Click **OK**.

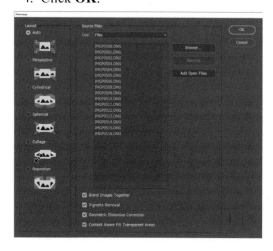

Now, take your hands off the keyboard and wait for Photoshop to do the work. You can work in a different window while Photoshop processes your photos, but don't try to interfere with Photoshop's automated action.

When it's done, if the panorama worked properly, you'll see the final result:

OK, that example didn't work well at all. However, I repeated the stitching process, selecting **Cylindrical** instead of **Auto** on the Photomerge dialog, and it produced a much better result:

Photoshop creates multiple layers for your image, each with a mask defining which parts of that image are being used in the final picture.

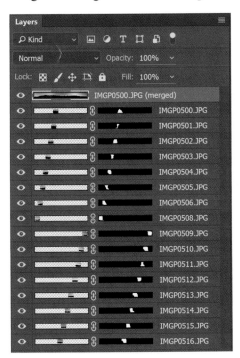

Understanding Panorama Layers

Photoshop creates a separate layer for every source image. If you select **Content Aware Fill Transparent Areas** on the Photomerge dialog, it adds a layer on top of the other layers and fills in the edges of the panorama.

If you hide all layers except for a single image layer, you can see which parts of that image Photoshop used in the final panorama, as demonstrated here:

The irregular edges are due to the layer mask. You can fill the layer mask with white to see how Photoshop altered the image prior to blending. For example, because I selected the **Correct Geometric Distortion** checkbox on the **Photomerge** dialog, Photoshop curved the straight image so that it would better fit in with other pictures.

Repairing Artifacts

Panoramas rarely turn out perfect. Before you publish a panorama, you should zoom into at least 1:1 and carefully examine each part of the image. Use these tips to fix common types of problems:

- **Bad Edges**. Where two pictures meet, the edges might not line up correctly. Typically, the fastest way to resolve this is to use the Spot Healing Brush and Content Aware Fill tools, as discussed in Chapter 3.

- **Poor Content Aware Fill.** When you select the **Content Aware Fill Transparent Areas** checkbox, Photoshop fabricates content to fill areas of the sky or foreground that you didn't capture. You might simply want to crop these areas out. After all, you didn't capture them in the original scene. Otherwise, you can manually clone or heal those areas until they look natural.

- **Moving subjects.** Often, a tree branch or object will have moved while you were taking pictures, and Photoshop will cut that moving object in half where two pictures meet. To fix this, find the layers that compose that part of the panorama and paint white in the layer mask of one of the images so that the entire moving subject is shown from a single exposure. Move the edited layer to the top.

Troubleshooting Failed Panoramas

Though it's more reliable than Lightroom, Photoshop is very picky about the images that it will process into a panorama. If it runs into any problem, it gives you a very vague error, or Photoshop might crash entirely.

If you do run into problems (and you certainly will), these tips can help:

- **Eliminate duplicate pictures**. You can't include two pictures of the same part of the scene; delete all but the sharpest image for any one part of the scene.
- **Divide the panorama into multiple sections.** If you're combining 8 pictures into a single image and you're getting an error, try stitching the left 4 pictures together separate from the right 4 pictures. You could then stitch the two panoramas together into a full image. Dividing the panorama will also help you identify which image is causing Photoshop to fail.
- **Use other software.** Microsoft ICE processes panoramas much more reliably, and it can be downloaded for free at *sdp.io/ice*. However, it will not process raw files, so you should do color correction, exposure adjustments, and highlight/shadow recovery prior to exporting the JPG files for use in ICE.

Or, it might be easier to export the images to JPG files and process them using Microsoft ICE. Here's the panorama that Microsoft ICE automatically produced with the same images. Note that ICE lacks Content Aware Fill, so there are black areas in the sky and foreground. However, it's easy to open the resulting file in Photoshop to fill in areas of the sky.

Speeding up Panorama Processing

Photoshop can take a long time to process a big panorama of raw pictures—anywhere from five minutes to hours, depending on the number of images and the resolution of each image. If you regularly create panoramas, this can be really frustrating (especially when Photoshop crashes after you've waited twenty minutes).

Unfortunately, there's not much you can do to improve the performance. Having 16GB or more of memory definitely helps, but the main performance bottleneck is your computer's processor. Photoshop uses only a single processor core, so if your computer has 16 cores (as mine does), Photoshop will use only 1/16th of your computer's processing capabilities. As a result, even a very fast computer will behave like a very slow computer.

14 HDR

Watch training videos at:
SDP.io/PSEDL

Dynamic range is the difference between the brightest and darkest parts of a scene. It happens to be one of the greatest challenges in photography because your eyes and brain can perceive over 20 stops of dynamic range in a scene, but a typical photo only shows 8 or 9 stops.

That's why, when you're talking to a friend outside in the shade, you can see his or her face and detail in the clouds. If you take a picture, however, either the clouds are completely overexposed or your friend's face is overexposed.

Many types of photography struggle with dynamic range, but it is especially challenging for landscapes, portraits, wildlife, and sports. In this chapter, I'll show you different techniques for using Photoshop to create images with dynamic ranges that better represent how you perceived the scene.

Recovering Highlights & Shadows from a Raw File

The simplest way to show massive amounts of dynamic range in a photo is to capture that image in raw, rather than JPG. Raw files contain all the information your camera's sensor gathered while taking the photo. JPG files, however, discard large amounts of shadow and highlight detail at either extreme of the dynamic range. For detailed instructions for capturing raw pictures with most common cameras, visit *sdp.io/tutorial*.

When you open a raw photo, Photoshop automatically launches Camera Raw. As shown next, the sample photo of the sun rising behind tufa formations at Mono Lake in California shows extreme dynamic range. The sky, lit by the rising sun, is a bit overexposed. The foreground is in complete shadow, creating a black silhouette.

It's an interesting photo, but the camera didn't capture the scene the way I saw it. To my eyes, the sky was far more interesting and colorful, and I could see the texture of the tufa in the foreground. Fortunately, because I shot raw, the file did capture the detail I saw in the scene; I just have to tease it out.

If you want to create images with that oversaturated, edge-glowing "HDR look," download Photomatix Pro from *sdp.io/photomatix*.

Using Camera Raw

Before you open a raw image in Photoshop, make adjustments to the Camera Raw sliders. These don't have to be your final adjustments, but you should try to get them as close as you can. Alternatively, you could do these same adjustments in Adobe Lightroom; that's generally easier.

I made these adjustments:

1. I lowered the **Highlights** and raised the **Shadows** as much as possible. This maximizes the dynamic range shown in the final image.
2. I adjusted the **Exposure** and **Contrast** until the sky looked nice.
3. I moved the **Whites** and **Blacks** sliders so that the picture had just a bit of clipping.
4. I clicked **Open Image**.

Fixing Chromatic Aberration

With the image open in Photoshop, I immediately noticed a problem common in high-contrast scenes: chromatic aberration that causes fringing of various colors, especially red, cyan, green, and magenta.

To fix that, select **Filter | Lens Correction**. This shows a 400% close-up of the tufa before fixing the fringing:

I selected the Custom tab and manually adjusted the **Chromatic Aberration** sliders to minimize the fringing. It's not perfect, but it's better. Though this is a Photoshop book, I find that Lightroom does a better job of automatically removing chromatic aberration without reducing the overall sharpness of the image. A more precise but time-consuming technique is to manually remove the fringing by sampling the fringe colors and then desaturating them for the edges in a new layer.

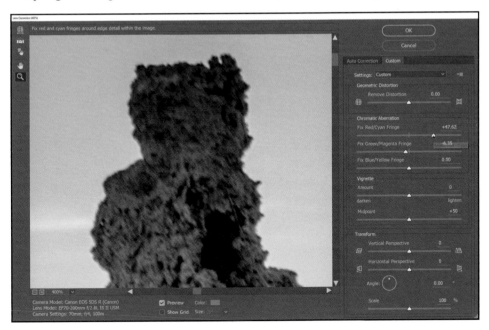

Restoring Contrast with Dodging & Burning

When you recover the shadows and highlights in a raw photo, you're cramming more dynamic range into a single image. Instead of a typical 8 stops of dynamic range, you're fitting 12 or more stops. Therefore, the entire scene has less contrast, and the final picture can look a bit washed out.

To solve that, create separate layers for different parts of the scene, so you can adjust the exposure of the foreground and background separately. For detailed information on how to separate the scene into layers, refer to Chapters 4, 5, and 6. In particular, you might find the **Select | Color Range** tool useful when set to **Highlights** or **Shadows**.

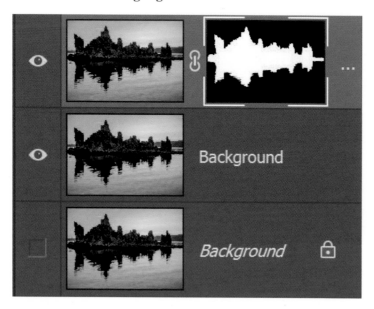

Next, I created a Levels adjustment layer and linked it to the Foreground layer by **Alt/Opt**-clicking the line between the new adjustment layer and the foreground layer. Then, I adjusted the levels in the adjustment layer to bring out the detail in the foreground, as shown next.

I repeated the process of adjusting the levels for the background layer.

Finally, I used dodging and burning to further increase the contrast. Here's my final edit, which more closely resembles the scene as I saw it:

The editing still isn't perfect; there's a bit of haloing between the foreground and background. That haloing occurs because of the extreme dynamic range recovery, and it's difficult to completely overcome without scaling back the processing. Fortunately, most untrained photographers don't notice the haloing.

Noise in Shadows

The less light your camera gathers for a part of a picture, the more noise the picture will have. Because they get less light, shadows always have more noise than highlights. However, you might not notice it, because the shadows are dark.

When you heavily recover shadows, as I did in the previous example, noise becomes more obvious. For example, look at a close-up of the foreground subject, which was completely black in the original photo. It appears speckled and lacks sharp detail.

The best way to overcome this is to change your photographic technique. Rather than capturing a single exposure, bracket three or more exposures, and then combine those bracketed exposures in Lightroom or Photoshop. For detailed information about bracketing, refer to Chapter 11 of *Stunning Digital Photography.*If you only have a single image to work with, you can apply noise reduction selectively to the shadow areas where noise is distracting. For detailed information, refer to Chapter 7, "Noise."

Bracketing & Blending Multiple Photos

If you bracket a scene, you can manually blend the multiple images together to show the best parts of both exposures. Consider these two sample images, taken at 1/20th (exposed for the sky) and 1/6th (exposed for the foreground).

To blend these images, follow these steps:

1. Open the images as layers in Photoshop.
2. Select both layers using **Select | All Layers**.
3. Select **Edit | Auto-Align Layers**. Even though I kept the camera on a tripod, the tripod shifted slightly, so the alignment helps.

4. Add a layer mask to the top layer.

5. Use the Gradient Tool (G) to paint a white-to-black gradient in the new layer mask, smoothly blending the two images together. For more information about masking, refer to Chapter 6.

The gradient tool precisely simulates the effect you'd get from using a graduated neutral density (Grad ND) filter at the time of shooting, but it allows you to more precisely control how rapidly the exposure transitions.

Longer gradients create more natural transitions. Experiment with different lengths of gradients to find your ideal.

Using HDR Pro

Photoshop also includes HDR Pro. I personally don't find it to be as useful as the techniques discussed earlier in this chapter, but I'll cover it for completeness. To use HDR Pro:

1. Select **File | Automate | Merge to HDR Pro**. Select the bracketed pictures you want to merge, select the **Attempt to Automatically Align Source Images** checkbox, and click **OK**.

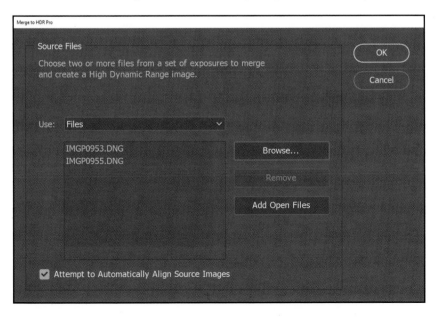

2. Adjust the settings as desired, and click **OK**.

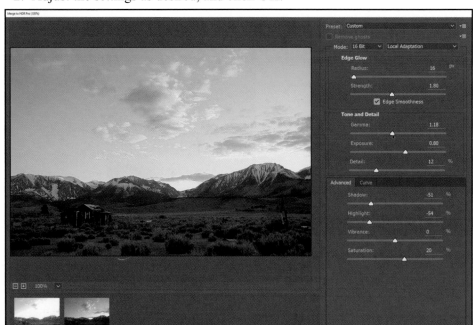

Here's an overview of the HDR Pro settings:

- **Preset**. Typically, I click this list and select the first setting. Then, I press the cursor down key on my keyboard to scroll through all the presets and select the preset that's closest to my desired result.

- **Remove ghosts**. If there was a movement in your picture (even just blowing leaves or grass) select this checkbox and HDR Pro will attempt to remove that movement.

- **Mode**. This controls the bits per pixel, with 8 bits providing 256 different gradients, 16 bits providing 65,656 gradients, and 32 bits providing over 4 billion gradients. Generally, the default is fine. The second list box is set to **Local Adaptation** by default, and that is usually the best value because it gives you the most control over the processing.

- **Edge Glow Radius**. Edge glow is an aesthetic trait of HDR processing created when the HDR software blends dark and light areas of different images. You can adjust the radius of the glow to change how far that glow reaches. The best looking value will vary depending on the subject and the megapixels in the image.

- **Edge Glow Strength**. This value controls how distinct the edge glow is. Adjust it to your own taste.

- **Gamma**. Gamma controls the overall contrast in the image. Lower values have higher contrast.

- **Exposure**. Exposure controls the overall image brightness.

- **Detail**. Detail adjusts the level of sharpening or softening in the image.

- **Shadow and Highlight.** These settings work exactly as they do in Adobe Lightroom, allowing you to recover the shadows or highlights to show more dynamic range in your photo. Often, your photo will look more natural if you lower the Shadow value or increase the Highlight value to show more contrast.

- **Vibrance and Saturation**. Just as they do in Adobe Lightroom, these values adjust the intensity of colors. Vibrance is a much more intelligent tool than Saturation, and you should usually leave Saturation at the default setting.

- **Curve**. This tool behaves exactly like the Curves adjustment layer, as described in Chapter 2.

I prefer to merge HDR images in Lightroom because the user interface is faster. To merge bracketed images in Lightroom, select them, right-click them, select **Photo Merge**, and then select **HDR**.

15 PUPPET WARP

Watch training videos at:
SDP.io/PSEDL

As a photographer, I find Puppet Warp useful for changing poses. For example, you could change the angle of an arm, leg, or wing—or an entire body.

At a high level, follow these steps to effectively use Puppet Warp:

1. Separate the object you want to warp from the background. You should have two layers: a background layer with the subject completely removed and a foreground layer with your subject masked. For detailed information, refer to Chapter 4, "Selections," Chapter 5, "Layers," and Chapter 6, "Masking."

2. Select the layer containing the image, and activate Puppet Warp using **Edit | Puppet Warp**.

3. Place pins in the parts of your subject that you don't want to move.

4. Place a single pin in the part of your subject that you want to move. Drag that pin to change your subject's pose.

The sections that follow will discuss these steps in more detail using the sample image available at sdp.io/psedl. The goal with that photo is to raise Madelyn's arms a bit higher, so it looks like she's balancing. The next examples show the before and after photos.

Separating the Foreground Subject

For best results, separate the subject you want to manipulate from the background. Using our sample image, I'll use the **Select | Select and Mask** tool to select Madelyn and output the selection as a new layer with a layer mask.

Notice that I have Shift Edge set to **+100%**; in this example, it's okay if I move a bit of the background with her, but it would look strange if parts of her didn't move. Therefore, I'm erring on the side of selecting too much.

Output the selection to a layer with a new layer mask, and then click **OK**.

Creating a Clean Background

With Madelyn on a separate layer, I need to create a background layer without Madelyn. That way, when I reposition Madelyn, the background will look normal.

1. Duplicate the background layer and name it **Background without Madelyn**.

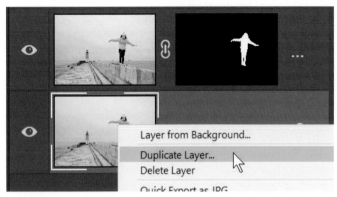

2. Because we created a new layer with a layer mask, copy the selection from the layer mask by right-clicking it and selecting **Add Mask To Selection**, shown next.

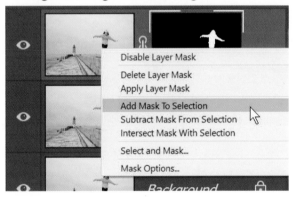

3. Now, use Content Aware Fill to remove Madelyn from the background. Select the **Background without Madelyn** layer, make sure the layer is visible, and then select **Edit | Fill**.

4. In the Fill dialog, select **Content-Aware** fill, as shown next.

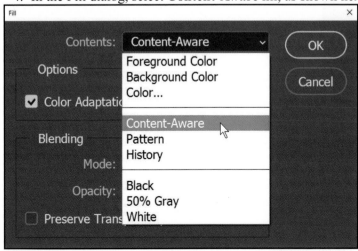

Photoshop removes Madelyn from the layer.

5. To see the background layer, hide the foreground layer. As you can see in the following example, Content-Aware Fill left a few artifacts.

6. Clear your current selection by pressing **Ctrl+D** or **Cmd+D**. Then, remove those artifacts using the Spot Healing Brush or Patch tools, as discussed in Chapter 3, "Healing, Cloning, Patching, & Moving."

7. Now, un-hide the foreground layer, and the image should look exactly like it did when we started. However, because we have separated the layers, we can freely pose Madelyn.

Moving the Subject

To use Puppet Warp to move the subject, follow these steps:

1. Select the foreground layer, and then select **Edit | Puppet Warp**. Puppet Warp draws a mesh over the masked part of the layer, as shown next.

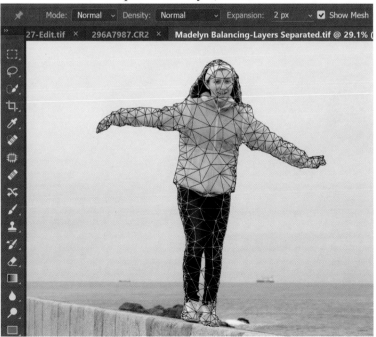

2. Click the **Density** list and select **More Points**. This just allows you to place more pins in close locations, generally without any penalty.

3. Now, click to place pins where you don't want the subject to move. I just want to raise her arms. Because I don't want her body or head to move significantly, I'll place pins to hold them still.

You don't have to place the pins precisely; just drop them roughly around her body. If you accidentally move a pin, or place a pin in the wrong spot, Alt/Opt-click it to remove it.

4. With her body held in place, drop a pin on each of her hands. Then, drag each hand pin up, as if she were holding her arms higher. To see the picture better, clear the **Show Mesh** checkbox.

5. Those pins caused her arms to curve unnaturally. If your subject has bones, drop and drag a few more pins to make the pose more natural.

6. When you're happy with the results, press **Enter** or click the check mark on the toolbar.

Rotating Around a Point

Dragging points, as described in the previous section, tends to create the most natural results. You can also rotate around a point by first clicking a point to select it and then holding the Alt/Opt key and dragging in a circle around a point (shown next).

Like other aspects of Puppet Warp, rotating around a point requires at least two points in the image. If you simply want to freely rotate an entire layer, use the **Edit | Free Transform** tool and use the center marker to control where the layer rotates, as shown next.

Using the Puppet Warp Toolbar

The previous example demonstrated the most important Puppet Warp features, but for completeness, I'll describe each of the tools on the toolbar.

Mode

Normal is good for most photography tasks. **Rigid** is slightly more strict about maintaining the structure of your subject, so if your puppet is getting too bendy, **Rigid** might work better. **Distort** allows Puppet Warp to change the size of your subject.

The following photos show the same warp performed with Rigid, Normal, and Distort modes.

Density

Adjusts the size of the mesh. Selecting **More Points** can make it easier to add points to your image, but it might slow your computer down.

Expansion

Controls how far outside your current selection or mask Puppet Warp draws the mesh. It's exactly like expanding or contracting your current selection. If you have used a layer mask, having a zero or positive value here will reduce artifacts.

Pin Depth

If you warp a subject enough to make it overlap, use the two Pin Depth toolbar buttons to determine which pins should be on top. Simply click a pin, and then click the Pin Depth up or down buttons.

The following example shows the difference between having the pins on the arms on top or beneath the face.

16 FIXING CAMERA SHAKE

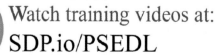

Watch training videos at:
SDP.io/PSEDL

If you hand-hold an image with a slow shutter speed, or you use a tripod and something shakes the tripod, the image will appear shaky. Photoshop has a tool that can minimize the appearance of camera shake, sensibly named Shake Reduction.

Fixing a Small Amount of Camera Shake

Here's an example landscape photo, zoomed in to show the shaky detail caused by the wind blowing the tripod:

To improve it, follow these steps:

1. Select **Filter | Sharpen | Shake Reduction**.

2. Optionally, adjust the **Blur Trace Bounds** to a value slightly bigger than the length of the movement in the shot. For example, if the camera shake only extends across 10 pixels, you can set the **Blur Trace Bounds** value to a low number, such as **15**. You can usually just leave this at the default.

3. Shake Reduction adds sharpening, and sharpening highlights the appearance of noise. If the noise becomes distracting, increase the **Smoothing** value. If no noise is visible, or you'd rather remove noise manually (as described in Chapter 7), reduce the **Smoothing** value until the noise almost becomes unpleasant.

4. Optionally, select the **Artifact Suppression** checkbox and set the **Artifact Suppression** value to **0**. You'll probably see overly harsh changes to your photo. Gradually increase the **Article Suppression** value until the picture looks natural.

5. Click **OK**.

The Shake Reduction tool isn't perfect, but it's better than nothing. It's slow, even with a fast computer. To improve the performance, clear the **Preview** checkbox. Then, use the smaller Detail pane to preview the effects of your current settings.

You can click the Undock Detail button in the lower-right corner and drag the Detail pane around the window to preview the effect of shake reduction on specific parts of the photo.

You'll always get better results by taking a steady picture. However, shake reduction can get usable results from an otherwise unusable picture. Here is the before and after of a tight close-up of the landscape photo:

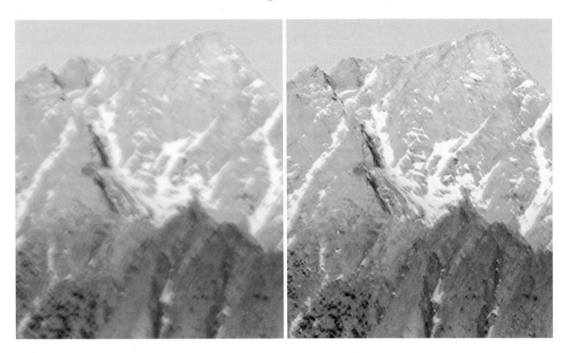

Fixing a Large Amount of Camera Shake

You can manually define the angle and length of the motion blur, too. Expand the **Advanced** section to see the automatically defined blur trace. You can right-click the blur trace to see what Photoshop based its correction on, as shown next.

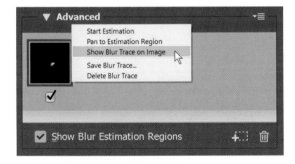

To manually specify the blur trace, clear the checkbox for the automatically defined blur trace. Then, press **R** or click the Blur Direction Tool in the upper-left corner.

Now, trace the motion blur. The easiest method is to zoom in close on a bright highlight in the photo, such as a reflection from something shiny. Click and drag across the motion, as shown next. This defines values for both **Blur Trace Length** and **Blur Trace Direction**.

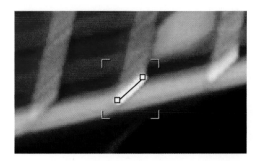

Now, you can fine-tune the processing by dragging the **Blur Trace Length** and **Artifact Suppression** sliders up or down. When the picture looks as good as it can, click **OK**. It'll never be perfect, but hopefully it's better than when you started.

With this example picture, no amount of tweaking produced a usable result. However, some parts of the image were better than before. The next examples show a close-up before and after processing.

17 ACTIONS

Watch training videos at:
SDP.io/PSEDL

Actions perform a series of steps that would otherwise take many clicks or several minutes. Photoshop has several actions built in, and you can download pre-made actions from the web.

Playing and recording actions is done from the Actions window (**Windows | Actions**). If you frequently use actions, you might want to leave it open and update your workspace (as described in Chapter 19, "Customizing Photoshop").

The Photoshop Actions system is a very primitive scripting language, and I don't want this chapter to become a long and complex study of programming logic. Therefore, I'm only going to show you the basics you need to know to download actions from the Internet, run and edit them, and create your own simple actions.

Playing Actions

To play an action, select the part of your image you want to apply the action to, open the Actions window, select the action, and then click the **Play** button.

Here's an example, showing how to add a water reflection to text on this chapter's sample image (available at *sdp.io/psedl*).

1. Add text to the image.

Carolina Wren

2. Select the layer with the text.
3. Optionally, use the History panel to make a snapshot. A single action creates many entries in your History, which might make it impossible to undo earlier changes.

4. In the Actions window, select the Water Reflection action and click **Play**.

5. Wait a moment while Photoshop plays the action. Some actions will happen in less than a second, while others might take several minutes.

In this example, the Water Reflection action did a great deal of complex work and added a new layer.

Using Button Mode

To simplify your view of the actions, turn on Button Mode by opening the Actions window menu and selecting **Button Mode**. As shown next, the actions appear as colored buttons.

Batch Processing Actions

You can run an action on multiple files, which can save a great deal of processing time. For example, if you want to add a watermark to 20 files, create an action that adds the watermark, and then use batch processing to apply the watermark to the files and save them in a new location.

Follow these steps to batch process any action:

1. Select **File | Automate | Batch**.

2. In the Batch dialog, shown next, select the **Set** and **Action** that you want to run. Note that you can only run one action. If you want to run multiple actions, create a new action that performs all the tasks of the multiple actions.

3. Click the **Source** list to choose to process the files you already have open in Photoshop, or all files in a folder. If you just want to process a handful of files, it's usually easiest to open them individually in Photoshop and set the source to **Opened Images**.

4. Click the **Destination** list to choose where to save the processed files. Select **Save and Close** to overwrite the existing files. To leave the existing files in their current state and saved the changes files with a new name or location, select **Folder**, and then specify the folder and filename to use.

If Photoshop normally prompts you for options when saving a file, it will interrupt the action. For example, when you save a JPG file, Photoshop prompts you to select the compression level. To avoid separately clicking **OK** for every file you process in the batch, save the file as the last step in your action, and choose not to save the file in the Batch dialog.

5. If you want Photoshop to continue processing files even if an error occurs, click the Errors list and select **Log Errors to File**. Otherwise, leave the default selection of **Stop for Errors**.

6. Click **OK**.

Photoshop will run the action for every file that you selected. If you're processing many files, the entire process can take hours.

Downloading and Installing Actions

Honestly, none of the built-in actions are especially useful. People do make really useful Photoshop actions, many of which can be downloaded for free from the Internet. Try searching for "Free Photoshop Actions."

Most actions you download will be in a file with a .ZIP extension. ZIP files must be extracted before they can be used. The exact steps vary depending on how your computer is set up, but usually you can right-click the ZIP file and click **Extract All** or something similar.

Within the ZIP file, you'll find an .ATN file. I suggest placing all the .ATN files you collect in a single folder on your computer, such as a folder named Documents\Photoshop Actions.

Once you've saved an .ATN file to your computer, install it in Photoshop by following these steps:

1. In the Actions window, click the menu icon in the upper-right corner, and then click **Load Actions**.

2. Select an action, and then click **Load**. You need to load each action separately.

You can create sets to organize actions by clicking the folder icon on the Actions window, as shown next. Many actions that you download will be nested within a set, and you must expand the set to play an action.

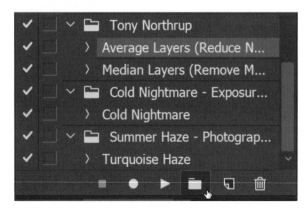

While some actions are complex and powerful, most of the actions that you will find on the Internet simply adjust the colors and vignetting to create a mood in the image. Typically, I prefer to use Lightroom Presets for that type of editing.

Recording Actions

If something takes more than a few seconds and you do it more than once a week, it's probably worth making an action. With an action, you can repeat a complex set of steps instantly.

To create a custom action, follow these steps:

1. If necessary, turn off button mode by clicking the Actions menu in the upper-right corner of the Actions pane and selecting **Button Mode**.

2. Optionally, create a new set to store your action. You can only share action sets, not individual actions, so you will almost always want to create a new action set.

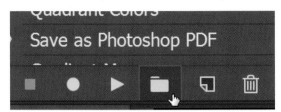

3. Click the **New Action** button on the Actions window.

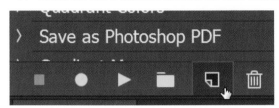

4. In the New Action dialog, type a name. You can place it within a set, assign a function key shortcut, and assign a color to make it easier to find in the future. Then, click **Record**. In this next example, I'm creating the Old Timey action, placing it inside the Tony Northrup set, and assigning the Shift-Control-F4 function key.

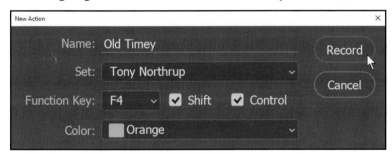

Photoshop automatically starts recording. Take your time; Photoshop only records the actions you perform within the app. At this point, you'll simply perform whatever steps you want to repeat in the future, and then click the Stop button. For the sake of this example, we'll do some stylized editing to the sample bird photo.

1. Make sure your photo layer is selected, and then select **Filter | Camera Raw Filter**.

2. On the **Basic** tab, set **Highlights** to **+50**, **Vibrance** to **+25** and Clarity to **+50**.

3. Select **Filter | Lens Corrections** and then select the **Custom** tab and set **Vignetting** to **-82**. Click **OK** to apply the Camera Raw Filter adjustments.

4. Select **Filter | Sharpen | Sharpen More**.

5. Click the **Stop** button on the **Actions** window.

That's it! You can now open any image, view the Actions window, select your action, and click Play. Photoshop will style the image in the exact same way.

After recording your action, you can add more steps by selecting the action and clicking the **Record** button. You can drag-and-drop steps to change their order. You can delete steps by selecting the step and then clicking the trash can button, as shown next.

Examining and Editing Actions

You can examine edit actions that you download or record. If you find an action you like on the Internet, looking through the steps is a great way to learn how to create the effect manually. You can also edit actions to tweak them slightly without needing to completely re-record them.

To view or edit an action, simply expand it in the Actions window. Photoshop shows you every step of the action. You can double-click a step to modify the values.

If you want Photoshop to pause after every step of the action, click the Action window menu (in the upper-right corner of the Action window), select **Playback Options**, and then choose **Step by Step** or **Pause For**, as shown next. Photoshop will briefly pause between each step, showing you exactly what is happening.

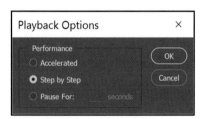

For example, I changed the vignetting amount in the previous step from -82 to -80 by double-clicking the Camera Raw Filter step and then using the dialog to change the vignetting amount.

To add steps to an action, select it and then click the **Record** button (shown next). Perform the steps you want to add to the action, and then click **Stop**.

Showing or Hiding Dialogs

By default, Photoshop hides windows that appeared when you initially recorded an action. That's generally useful. However, if you want to make changes to a dialog during the action, select the dialog box for that step, as shown next.

When you play the action, Photoshop will show that dialog with the settings configured in the action. You'll be able to customize those settings before the action continues.

If there's an existing action that shows a dialog and you don't want it to, you can clear the dialog setting.

Showing Messages

You can use stops to show a message to the user. Click the **Action** window menu (the box with four lines in the upper-right corner) and select **Insert Stop**. Type your message, and select the **Allow Continue** checkbox.

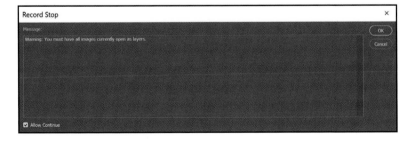

When you run the action, a dialog such as the following will appear at that step.

Using Conditionals

If you've used programming or scripting languages in the past, you're familiar with conditionals such as if-then-else. Conditionals check the state of some aspect of the image while the action is running and then run different actions depending on whether the value is true or false.

To add a conditional, click the **Action** window menu and select **Insert Conditional**. As shown next, you'll select one of the very limited set of conditions you can check. Then, you'll select separate actions that should run if the condition is true or false.

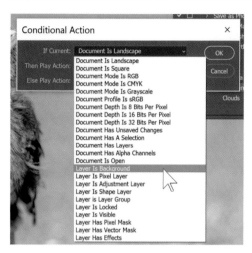

As an example, let's create a simple condition that checks whether the current layer is the background layer. If it is, we'll inform the user that he or she can't run the action on a background layer, and we'll stop processing.

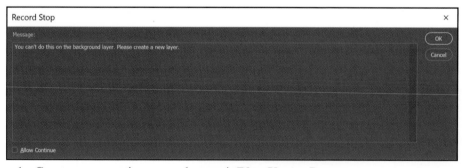

1. Create a new action set and name it **Blur Upper Layer**.

2. Within that action set, create a new action and name it **--Not on the Background**. The two hyphens before the action name help to indicate to the user that he or she won't run it directly.

3. Click **Record** to start recording the action.

4. Click the Action window menu and select **Insert Stop**.

5. Type a message informing the user that he or she can't do that on the background layer.

6. Stop recording that action. Users won't ever run that action directly; it simply displays a message and stops processing further steps.

7. Create a new action in your action set and name it **RUN ME: Blur Upper Layer**.

8. Click the Action window menu and select **Insert Conditional**. Click **OK**.

9. Create a condition that checks if the current layer is the background layer. If it is, run the **Not on the Background** action, as shown next.

10. Select **Filter | Blur | Blur More**. This step will only run if the layer is not the background layer.

11. Stop recording the action.

12. Move the **RUN ME** action to the top of the action set so that it's easier to find.

Your action should now resemble the following figure.

Now, run the **RUN ME: Blur Upper Layer** action on the background layer. The action will display the message you created and then stop running.

Create a new layer, and run the action on the new layer. The action will not display the message because it only appears if the current layer is the background layer. The action will skip the final step (the Blur More filter).

This is a very simple example of how to run different processes based on the current state of the image. You can certainly get more complex. For example, if the action detects that the current layer is the background layer, it could duplicate it and then continue the action.

Sharing Actions

To share an action set that you've created, select the action set in the Actions window, click the Actions window menu, and select **Save Actions**. Then, save the .ATN file to a folder. You can share the .ATN file with other people.

To load your action set on another computer, click the Actions window menu and then select **Load Actions**.

18 UNDERSTANDING IMAGE FILES

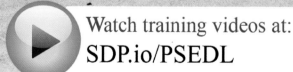

Watch training videos at:
SDP.io/PSEDL

Before we get into details, here's what you really need to know about files:

- Save most edited files as a .PSD.
- If the file is larger than 30,000 pixels horizontally or vertically, Photoshop will tell you to save it as a .PSB.
- If you're opening the file in a non-Adobe app after editing, save it as a .TIF. Use .ZIP image and layer compression.
- If you're printing or sharing a picture, save it as a full-resolution .JPG file with 80% quality.
- Don't worry about the DPI or PPI; just save and share all images in their native resolution.

File Types

Here's an overview of the most important file types you'll work with in Photoshop.

Format	Works with Browsers	Lossy Compression	Lossless Compression	Stores Layers	Supports Transparency	Supports Animation
JPG	✓	✓				
PSD			✓	✓	✓	
PSB			✓	✓	✓	
TIF		✓	✓	✓	✓	
PNG	✓		✓		✓	
GIF	✓		✓		✓	✓
DNG			✓			

JPG

JPG files are how you share images, whether you're uploading them to Facebook, sending images to clients, or having prints made by an online service.

The reason JPG is by far the most popular format for sharing pictures is that it's compressed. If you export a 20MB raw file as a JPG file, it might only be 20KB—1/1000th the original size.

Or, the resulting file might be 2MB, or 2KB. You can choose how much compression you use on your JPG files. Higher levels of compression produce smaller files, which are faster to upload and download. However, higher compression levels create lower-quality images, which might be less sharp or have visible blocks in them.

Compression makes big files smaller. There are two types of compression:

- **Lossless**. Lossless compression keeps the entire original file intact, reducing file size by a small amount, such as 5-25%. For example, if you take a picture at night and the entire sky is perfectly black, a losslessly compressed file might describe a large portion of the sky as black, rather than separately listing 4 million black pixels. TIF and DNG files offer lossless compression, and there's no reason not to use lossless compression when it's available.

- **Lossy**. Lossy compression can reduce the file size much more than lossless compression. For example, if your photo includes a large section of blue sky, lossy compression might describe a big block of the sky as a single particular shade of blue, even though it gets slightly brighter across the sky.

JPG files have a bad reputation, mostly because of compression artifacts. However, if you save your JPG file with reasonable quality, such as 75%, you probably won't notice any artifacts. There's rarely a reason to save at 100% quality.

Here are examples of the sky in a file saved at 0%, 25%, 50%, 75%, and 100%:

If you look at the first image (0% quality), you can probably see banding in the sky. The high compression described blocks of the blue sky as being the same brightness, so the gradient was no longer smooth.

Whether you can still see those gradients at 25% and 50% depends on how you're viewing the image. On my screen, as I'm writing this, the JPG image at 50% quality looks noticeably smoother than 25% quality, even without zooming in.

Viewing the images full screen on my 30" monitors, I can't see any difference between 50%, 75%, and 100% quality. If I zoom into 100% (1:1) to view individual pixels, and I look very closely, I can see that 75% quality looks slightly better than 50% quality in some parts of the picture.

No matter how closely I look, I can't see any difference between 75% and 100% quality, even though I chose this picture specifically because the smooth sky gradient would exaggerate differences. I also can't see any difference between 75% quality and the original raw image.

In summary, you will probably never need to use higher than 75% JPG quality, and you don't need to stress about sharing or printing JPG files.

The best way to save JPG files for sharing is to use the Export As dialog (**File | Export | Export As**). If you regularly export files with the same settings, set your defaults using **File | Export | Export Preferences** and then setting the **Quick Export Format** and **Quick Export Location** values, as shown next.

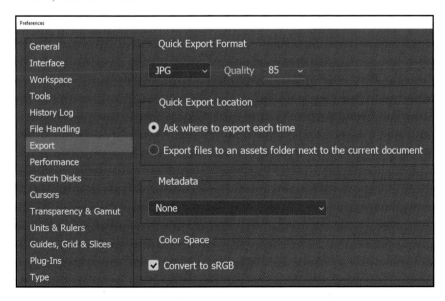

After setting your preferences, you can export images by choosing **File | Export | Quick Export as JPG**.

You should not save images as JPG files for future editing. Each time you open and resave a JPG file, the image quality degrades a tiny amount. TIF or PSD files do not degrade (with their default settings) and also save layer information, so those formats are better choices for editing.

PSD

For many photographers, PSD files are the best way to save images after editing them in Photoshop. PSD store layers and layer masks and don't degrade your image quality like JPG files. Therefore, you can save your edited image as a PSD, open it later, and see the same collection of layers.

About the only part of your image that isn't stored in a PSD file is your history: If you reload a PSD file, you won't be able to Undo tasks or access snapshots that you created in your History. This limitation of the file format is why it's important to use layers whenever possible. You can only use your History and snapshots during a single working session.

You shouldn't use PSD files for most online sharing or printing. However, if you're working with an editor or designer who uses Photoshop products, you can send him or her a PSD file and know he or she will be able to work with it.

Tip: If your PSD, PSB, or TIF files are getting too large, use these tips to minimize their size:

- Delete any layers that you know you won't later edit.
- Apply any layer masks that you know you won't later edit. To apply a layer mask, right-click the mask and then click **Apply Layer Mask**. This makes the layer partially transparent and reduces the total file size.
- If you have layers that extend beyond the canvas and you don't plan to resize or move those layers, crop the image to delete the unused image. Press **Ctrl-A**, then **Image | Crop**.

PSB

The PSB format allows you to save larger files than PSD. Otherwise, it's exactly the same as using PSD.

PSD files are limited to 30,000 x 30,000 pixels (about 900 megapixels). PSB files are limited to 300,000 x 300,000 pixels (about 90 gigapixels).

900 megapixels is pretty big, so you might never run into the PSD file size limitation. However, if you make a big panorama, it's easy to create an image that's more than 30,000 pixels wide. A 180-megapixel panorama in a wide format can easily be large enough that you can't save the image as a PSD.

So, why not use PSB for everything? Compatibility. Adobe Lightroom, and many other apps, can't read PSB files.

You don't need to memorize the PSD and PSB file limitations—if you make a panorama that's too large for the PSD format, Photoshop will automatically suggest you save it as a PSB file.

TIF

TIF (also known as TIFF) files provide the same benefits as PSD files: they're lossless and they store layers. TIF files are sometimes larger than PSD files and sometimes smaller. However, in my testing, the average PSD file size was smaller, so use PSD files whenever possible.

The only time you'll need to use a TIF file instead of a PSD file is when you need to open the file in an app that doesn't support PSD files. However, TIF is an open file format, whereas Adobe charges licensing fees for using the PSD format. Therefore, non-Adobe apps are unlikely to open PSD files. If you work primarily in Photoshop, that might never be an issue for you.

When you save a TIF file, Photoshop will prompt you to save layers (which is usually a good idea) and then show you the options shown in the next example. The defaults will always be fine, and most of the options haven't been required for at least a decade:

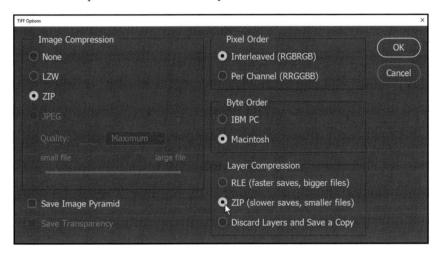

Here's an overview of these options:

- **Image Compression**. Typically, I select LZW compression, which is lossless. ZIP Compression is also lossless, and slightly more effective, but some outside apps don't support ZIP compression, and the storage benefit is less than 1% in my testing. JPEG compression is lossy, so it will reduce the file size further, but image quality will degrade as you edit a single image multiple times.

- **Save Image Pyramid**. You'll probably never need to select this checkbox. This option stores multiple resolutions of a single image, a feature required for some apps in the past, but modern photographers will probably never need it.

- **Save Transparency**. Likewise, you'll probably never need to select this checkbox, even if you have transparency and masks in your image. It's for backwards compatibility with really old apps.

- **Pixel Order**. This option changes how Photoshop stores the file on disk but doesn't change your image in any way. Technically, **Per Channel** might open files a bit faster and might compress them a tiny bit more, but you'll probably never notice a difference.

- **Byte Order**. All modern Mac and PC apps can read either byte order, so you don't need to change this option.

- **Layer Compression**. This option actually does matter: Selecting **ZIP** reduced the size of test files with one or more layers by 35-40%. Files were still much larger than the equivalent PSD files, however.

PNG

The PNG image format has one use: it can save a partially transparent image. This is useful for creating a cut-out picture that you add to a web page so that the subject floats above the background, even if the background changes. If you use video editing software, you can add a PNG file with transparency to a video so that the video shows through the edges of the image.

To save a transparent PNG, create a transparent image by creating a layer with a layer mask and a hidden background, as shown next.

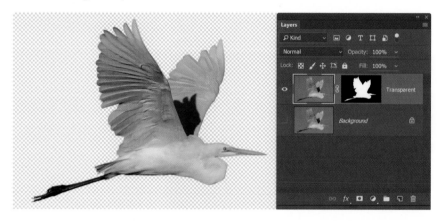

Then, select **File | Export | Export As**. In the Export As dialog, set the **Format** to **PNG** and select the **Transparency** checkbox. To reduce the file size, adjust the **Image Size** settings to exactly what is required by the website or video. Selecting the **Smaller File** checkbox will also reduce file size, but might reduce image quality, depending on the number of colors in your picture. Finally, click **Export All**.

If you embed a transparent PNG file into a web page, the background shows through. For example, the following bit of HTML shows an image with a light blue background:

```
<p style="background-color:lightblue;">
  <img src="http://northrup.photo/downloads/TransparentBird.png"/>
</p>
```

Without changing the image, a web designer could alter the color of the background to be any color or even another image. For example, this HTML changes the background to orange:

```
<p style="background-color:orange;">
  <img src="http://northrup.photo/downloads/TransparentBird.png"/>
</p>
```

GIF

Like PNG, the Graphics Interchange Format supports transparency. However, PNG files are usually a better choice, because PNG files are higher quality.

Note: The creators of the GIF format pronounced it with a soft G, as "jif." They'd joke, "Choosy developers choose GIF," copying commercials for a popular American brand of peanut butter. Many people pronounce it using the hard G sound from *Graphics*. Whichever pronunciation you choose, you will talk to people who will argue with you that you're pronouncing it wrong. Now, at least, you know both sides of the argument.

The GIF format has one unique trait: it supports animations, which are short videos without sound. To use Photoshop to make an animated GIF, follow these steps:

1. Open all the pictures as separate layers in Photoshop. For example, you might select a sequence of photos in Lightroom, right-click them, and then select **Edit In | Open as Layers in Photoshop**.

2. Select **Window | Timeline**. The Timeline window will appear, which you will use to create your video.

3. In the Timeline window, click the list and select **Create Frame Animation** (shown next).

4. Click the **Create Frame Animation** button. You'll see a single image on your timeline now.

5. Select all layers (**Select | All Layers** or **Alt+Ctrl+A**).

6. In the upper-right corner of the Timeline window, click the menu icon, and then select **Create New Layer for Each New Frame** (shown next).

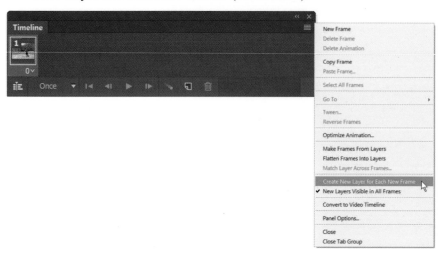

7. Click the same menu icon again, and select **Make Frames From Layers**. Photoshop adds every picture as a separate frame for every layer.

8. Now, test your video by clicking the play button on the bottom of the Timeline window, as shown next.

9. If the video plays backwards, click the Timeline menu and then click **Reverse Frames**, as shown next.

10. Most animated GIFs loop indefinitely. Click the drop-list at the bottom of the Timeline menu (it says **Once** by default) and then click **Forever**.

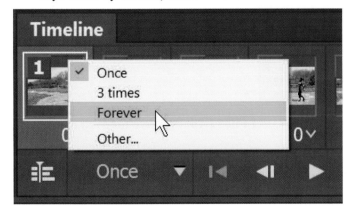

11. Now, calculate the correct delay between frames. To determine the correct speed, divide 1 by your camera's frames per second. For example, the sample pictures in this example were shot with a Canon 1DX Mk II at 14 frames per second. 1/14 = 0.07. If your camera shoots 5 frames per second, your delay will be 0.2, because 1/5 = 0.2.

12. Click the Timeline menu again and choose **Select All Frames**.

13. Set the delay between frames by clicking the drop-list below any frame (it shows **0** by default) and then clicking **Other**.

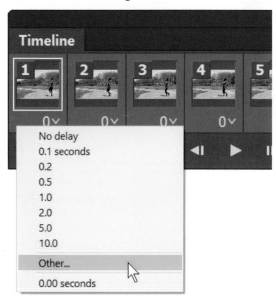

14. In the Set Frame Delay dialog, type the delay you calculated in the earlier step, such as **0.07** or **0.2**. Click **OK**.

15. Now, you're ready to save your GIF animation. Select **File | Export | Save for Web (Legacy)**.

16. In the **Save for Web** dialog, first adjust the **Image Size** to the size you need to share. Animated GIFs on the web aren't compressed like video files, so even short animations get very large. As a result, you should choose a small size to make sharing easier, like 600 pixels wide. Click and drag the **Percent** label to quickly adjust the size.

17. In the **Preset** list, select **GIF 32 Dithered**. This creates an animated GIF with 32 different colors. If your animation needs more colors to look good, you can increase it in the next step.

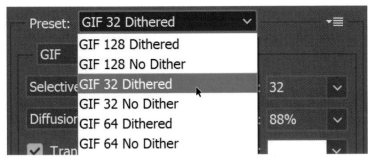

18. Click the Play button at the bottom-right corner of the **Save for Web** dialog. If your image quality looks terrible, click **Preset** list again and select **GIF 64 Dithered**. If it still looks bad, select **GIF 128 Dithered** from the preset list.

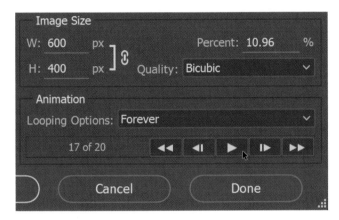

As you choose more colors, your animated GIF file size will increase. You can see the current file size in the lower-left corner of the Save for Web dialog.

19. Click **Save** to save your animated GIF.

Just because you can use Photoshop to make animated GIFs doesn't mean it's the best way to do it. Many websites will create animated GIFs from JPG images that you upload, including *gifmaker.me*.

Raw Files (DNG, CR2, ARW, NEF, etc.)

Most cameras can take pictures in either JPG or Raw format. Raw files store all the information captured by the camera's sensor, whereas JPG files discard a great deal of information to create a smaller and more easily shareable image. Using Adobe Camera Raw, Photoshop can open most common raw file types.

To save the image after editing it, use a PSD, PSB, or TIF file. You can't save edited raw images using a raw file format. This helps protect your original image so you don't accidentally mess it up.

Metadata

Most image file types contain metadata, which is text and numbers stored in the file format that don't change how the picture looks. For example, in Windows, right-click an image, select **Properties**, and then select the **Details** tab. On most pictures, you can see the details of the camera and settings, as shown next.

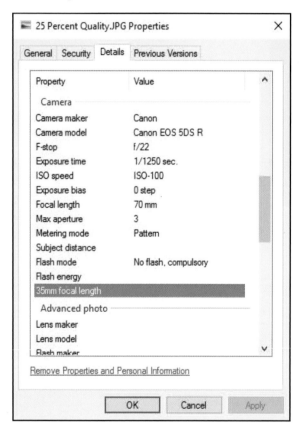

You can also use the Metadata panel in the Library view of Lightroom.

JPG, DNG, and TIF files tend to contain metadata, whereas PSD, PSB, and PNG files usually don't have camera metadata.

Metadata is overall useful; Lightroom and Photoshop can use it to automatically improve image quality based on the camera and lens that you used. When combining bracketed images for HDR, both apps can use the exposure information to properly blend images. Metadata can also be useful for finding pictures taken with a specific camera or lens, or with specific camera settings.

However, metadata can also be harmful. For example, many cameras store the GPS location with an image. If you were to share that image online, a creep on the Internet might use the GPS data and show up at your house.

When you share images online, be aware of the metadata that you're sharing. If you use the Export As dialog (**File | Export | Export As**), Photoshop will automatically remove all metadata except for your copyright and contact information. Alternatively, as shown next, you can choose to remove all metadata.

It's too bad that the **Export As** tool doesn't give you the option to select which metadata to share. When uploading images to sites like 500px or Flickr, it's nice to include your camera settings, so that other people can learn from that information.

If you do want to export your metadata, use **File | Save As** instead of the Export As tool. Note, however, that Save As keeps all your metadata intact, including data that might be personal.

Lightroom provides more powerful export tools, and gives you more control over your metadata. As a result, I usually close my image in Photoshop, find it in Lightroom, and then share it directly from Lightroom. As shown next, the **Metadata** section of Lightroom's export tool allows you to easily remove location info. Lightroom also makes it easy to automatically add your copyright information when you import new photos.

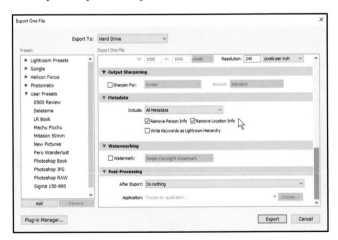

Color Spaces

Color spaces define how color and brightness information is stored in a file. I'll describe the two most common color spaces, RGB and CMYK.

RGB

Most images are stored in RGB (Red, Green, Blue) format. In other words, each picture is actually three separate pictures: red, green, and blue.

This makes sense, because the human eye has separate red, green, and blue color receptors (essentially), and most cameras have separate red, green, and blue pixels.

Any spot in the image that has a high value for red, green, and blue will appear white. Any spot that has a low value for red, green, and blue will appear black.

If a particular pixel has a high value for blue, but a zero value for red and green, it will appear as primary blue. Areas with a bit of red and blue, but no green, will appear purple.

In this way, displays blend red, green, and blue to simulate the entire visible spectrum.

CMYK

While most photographers work with the RGB color space, many printers and designers need to work in the CMYK (Cyan, Magenta, Yellow, and Black) color space. While RGB is designed to work with light, where a zero value represents black, CMYK is designed to work with paper and ink, where a zero value represents white.

You may never need to work with CMYK, but if a designer asks, you can switch from RGB to CMYK using **Image | Mode | CMYK Color**.

Bits per Channel

Each picture is made up of separate channels. For example, RGB images have separate red, blue, and green channels.

Each channel is made up of millions of individual pixels. Each pixel has a value. In the RGB color space, a value of 0 represents black for that channel.

The value that represents white can be either 255 (8-bit), 65535 (16-bit), or 4,294,967,295 (32-bit).

Note: A bit is a power of 2; that's how computers do math. $2^8 = 256$, $2^{16} = 65,536$, and $2^{32} = 4,4294,967,296$. Computers start counting at 0 instead of 1, so the maximum values are one less than the possible values for a given bit depth.

Having more values between white and black makes smoother gradients, because there are more possible values. However, every image you see online is stored in an 8-bit format, and people never complain about obvious jumps in gradients.

Nonetheless, 16-bit images can produce noticeably better results when heavily editing an image. For example, imagine that you edit a washed-out 8-bit image. You decide to use the Levels tool to raise the black point from 0 to 64, and lower the white point from 255 to 191. Now, instead of 256 possible values, you only have 128 possible values, and any jumps in smooth gradients are twice as obvious.

If you were editing a 16-bit image, even after adjusting the levels in the same way, you would still have 32,768 possible gradients for each color channel. With that many values, your eyes would never see any obvious jump.

The following two strips of sky are taken from the same picture of the flying bird. The first is 8-bits, and the second is 16-bits.

If you look closely at the two strips of sky, they are very slightly different, but one isn't obviously better than the other, even with editing more severe than any photographer would ever publish. The noise in the sky, even taken at ISO 100 with a full-frame camera, is far more prominent than artifacts created by limited bit depth.

Simply to demonstrate the theoretical benefit of higher bits per channel, I drew a very slight gradient in Photoshop, using just 5% of the entire range of brightness. Then, I added a Levels adjustment layer to expand that 5% gradient across the entire range from black to white. The next two images show how the images appears in 8-bit and 16-bit color depths.

You can see the difference in that example, but only because the I dropped about 5-bits of brightness information using editing. Thus, the 8-bit example actually shows only about 3-bits of information, while the 16-bit example still shows 11-bits of information.

From this we can conclude that 16-bits is better than 8-bits for editing, but that you'll probably never notice the difference in a published photo, because noise in photos will always be more obvious than artifacts produced by even the heaviest editing techniques.

Note: If you examine the metadata of pictures, you'll often see the bit depth listed as 24. RGB images have 3 channels, and each channel has 8 bits. The bits per channel is 8, but the total bit depth is 24.

Dots per Inch (DPI)/Pixels per Inch (PPI)

As a photographer, you don't have to worry about setting the DPI or PPI for an image. Generally, you should provide the full image resolution to your designer or printer, and let him or her (or the software) scale the image as needed.

In a nutshell, don't worry about DPI and PPI. However, many photographers have questions about DPI and PPI, so I'll describe them.

Dots per Inch (DPI) is the number of ink dots a printer makes in a straight line while printing a picture. If a printer had 2 DPI, the printed image would be extremely blocky, like you see when playing Minecraft. If a printer had 10,000 DPI, the printed image would be extremely sharp, even under heavy magnification.

Pixels per Inch (PPI) is the same concept, except that it refers to pixels on a monitor or other electronic display. On most displays, a pixel is either red, blue, or green, and adjacent pixels visually blend together unless you're really close to the display.

Most people can't perceive more than 300 DPI or 300 PPI when they put their face as close as possible to paper or a display. If you used a magnifying glass or reading glasses, or if you have better-than-normal vision, you might be able to see more detail. This 300 DPI/PPI limit is commonly accepted, so you'll often hear it brought up when photographers and designers discuss printing.

In Photoshop, you can use the Image Size tool (**Image | Image Size**) tool to adjust the width, height, and PPI (labeled Pixels/Inch) of an image. As long as the Resample checkbox is not selected, changing these values won't impact your image in any visible way.

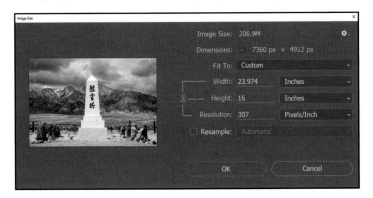

You could change the width to 1 inch, and then the resolution would be 7,360 Pixels/Inch. Or, you could change the width to 1,000 inches, and the resolution would be 7 Pixels/Inch. Either way, your picture is still the exact same picture. In practice, none of this matters at all.

The only place you'll see the PPI of the image is in the metadata of the file after you save it. If your image metadata has a resolution of 100 DPI, it doesn't stop it from being printed at 300 DPI, or any other DPI. Apps can completely ignore the metadata.

Many times in my career, I've had a client ask me for a specific image at 300 DPI. I could use the **Image Size** dialog to set the resolution to 300 Pixels/Inch, but that would be pointless unless I also knew the width and height of the printed image.

If a client wants an image at 300 DPI, you could ask him or her for the width and height that he or she is printing at, and then use the **Image Size** dialog box with the **Resample** checkbox selected to resize the image. However, there's no benefit to this, regardless of the format. It's always better to bring the full-resolution image into the design or printing software, which will automatically scale the image as needed. If you were to resample the image, and the designer changed his or her mind about the size of the image, he or she would need to ask you to reopen and resize the image again.

So, I usually just tell the client that the image I provided is at the full resolution, and it is 300 DPI. As I said at the beginning of this section, photographers really don't have to worry about DPI and PPI. Just get the sharpest, highest resolution image you can.

It's not related to Photoshop, but if you're interested in the relationship between DPI/PPI and the megapixels of your camera's sensor, watch the video at *sdp.io/mp*.

19 CUSTOMIZING PHOTOSHOP

Photoshop's user interface can be confusing and overly complex. As a photographer you will find that there are probably many tools and windows taking up screen space that you'll never need. There are probably other tools that you use regularly, but they're hidden or they require multiple keypresses to activate.

Fortunately, you can customize Photoshop precisely to your needs, which can make your editing both faster and more fun. This chapter covers the most important ways to customize your Photoshop environment.

Customizing Keyboard Shortcuts

Photoshop includes a dizzying array of keyboard shortcuts. Yet, there are still functions that I use on a daily basis that lack keyboard shortcuts.

For example, I often switch between the Healing Brush Tool, the Spot Healing Brush Tool, and the Patch tool. By default, all three tools are assigned the **J** keyboard shortcut. If I want to switch between them, I need to press **Shift+J** multiple times until Photoshop selects the right tool.

For the way I edit photos, it would be more efficient if each tool had its own keyboard shortcut. However, Photoshop only allows you to assign A-Z for tools, and you can't use **Alt**, **Opt**, **Ctrl**, or **Shift** as modifiers. Photoshop already has every letter of the alphabet assigned to existing tools. Therefore, if I want to assign shortcuts to two of my favorite tools, I need to remove shortcuts from existing tools.

I never use the default H keyboard shortcut, which selects the Hand tool. Instead, I hold down the space bar when I need the hand tool. Therefore, I can delete the H keyboard short and then assign it to the Healing Brush tool.

Assigning Keyboard Shortcuts to Tools

To customize the healing tools, follow these steps:

1. Select **Edit | Keyboard Shortcuts**.
2. In the **Shortcuts For** list, select **Tools**.

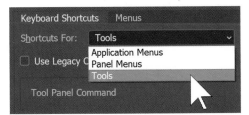

3. Select the Hand tool, and then select **Delete Shortcut**, as shown next.

4. Select the **Healing Brush Tool**, and then press **H** to assign that as the keyboard shortcut, as shown next.

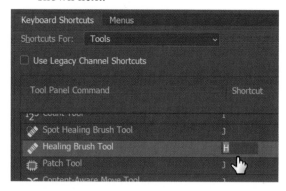

5. Click **OK**.

For the Patch tool, I decided to reuse the **K** keyboard shortcut, which is currently assigned to the rarely-used **Slice** tool.

If multiple tools have the same keyboard shortcut, you will need to hold down the Shift key while pressing the keyboard shortcut to alternate between them.

Assigning Keyboard Shortcuts to Menu Items

To assign keyboard shortcuts to menu items, use the **Edit | Keyboard Shortcuts** command, exactly as you did in the previous section. In the **Shortcuts For** list, select **Application Menus** or **Panel Menus**, as shown next.

Now, select the menu item that you want to assign a keyboard shortcut to, and press the shortcut. You can use the **Ctrl**, **Alt/Opt**, and **Shift** keys to create complex keyboard shortcuts for menu items.

For example, I assigned **Alt+Shift+Ctrl+A** to the frequently-used **Auto-Align Layers** command. As shown next, that keyboard shortcut was already in use by a command that I never access. Photoshop automatically removed the previous shortcut.

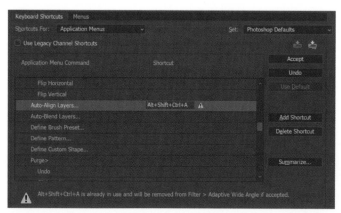

Afterwards, click one of the two buttons (shown next) to save your settings, and click **OK**.

Saving and Loading Keyboard Shortcuts

If you do customize your keyboard shortcuts, you'll find it extremely confusing to use a different computer. Fortunately, you can export and import your keyboard shortcuts. Unfortunately, Photoshop doesn't do this for you automatically, even if you login with Adobe Creative Cloud.

To save your keyboard shortcuts, click the **Create a New Set** button in the **Keyboard Shortcuts and Menus** dialog, as shown next. Type a name for your keyboard shortcuts, and then click **Save**. The button just to the left will save your changes to the current Keyboard Shortcut set.

Notice that you can create multiple sets of keyboard shortcuts. If two users share a single computer, you could each have your own keyboard shortcut sets and have each be part of your own workspaces. As shown next, be sure that you select Keyboard Shortcuts when creating a new workspace using **Window | Workspace | New Workspace**.

Customizing Menus

Photoshop menus are vast and complex, and most photographers won't need most of the menu items. You can hide menu items that you don't need to make those you do need easier to find. You can also change the color of menu items to highlight them.

As an example of how customizing menus can make it easier to find those commands you most frequently use, the following example shows the default **Edit** menu (with 35 items) and my customized **Edit** menu (with 11 color-coded items). Which menu makes it easier to find the frequently used **Auto-Align Layers** command?

Hiding Menu Items

After you've spent a few months working with Photoshop and have gotten familiar with the commands you regularly use, I recommend hiding commands that you've never used. Simplifying the menus makes finding commands much easier, and it can significantly speed your work. If you do need a hidden command in the future, it's easy to show it.

To hide a menu item, follow these steps:

1. Select **Edit | Menus**.
2. Expand the menu to find the command you want to hide.
3. Click the Visibility eye to hide the menu item. The next example shows me hiding the **Browse in Bridge** menu item, which I never use, because I don't use Adobe Bridge.

4. Click the **Create a new set button** (shown next), and save your customized menu with a unique name. If you have already created a custom set, click the button to the left to overwrite your existing settings.

5. Click **OK**.

If you need to access a hidden menu item, click **Show All Menu Items** at the bottom of the menu.

You can use keyboard shortcuts to access hidden commands. Therefore, if you always use a keyboard shortcut, you shouldn't hesitate to hide the corresponding menu item. For example, I always press **Ctrl+C/Cmd+C** to copy, and **Ctrl+V/Cmd+V** to paste. Therefore, I hid the **Edit | Copy** and **Edit | Paste** menu items to simplify that menu.

Adding Color to Menu Items

To add color to a menu item, follow these steps:

1. Select **Edit | Menus**.
2. Expand the menu to find the command you want to colorize.
3. Click the Color list for that menu item, and select the color you want to use, as shown next.

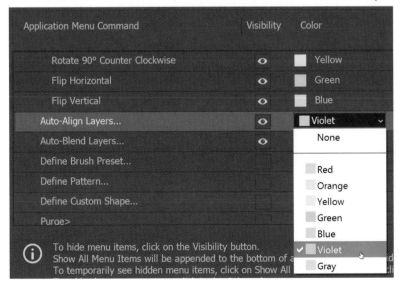

4. Click the **Create a new set button** (shown next), and save your customized menu with a unique name. If you have already created a custom set, click the button to the left to overwrite your existing settings.

5. Click **OK**.

I color code my most frequently accessed commands simply to make them easier to find. There's no particular system to my color coding; I simply assign random colors. Over time, I've learned to look for that specific color when accessing the command.

Customizing the Toolbar

You can customize the Toolbar using **Edit | Toolbar**, as shown next.

Typically, I use this to hide tools that I never use. For example, I never use the **Single Row Marquee Tool**, so I dragged it to the **Extra Tools** column. The next figure shows some of the extra tools that I removed.

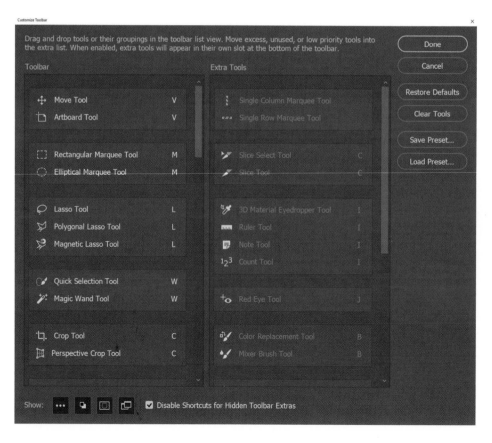

I also use it to bring frequently accessed tools to the top so I don't have to click-and-hold a button to access hidden tools. For example, I constantly access the different healing tools, so I made them all top-level tools.

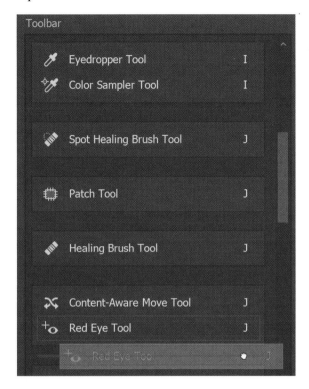

If you always use a keyboard shortcut to access a tool, you can hide it to save space. However, be sure to clear the **Disable Shortcuts for Hidden Toolbar Extras** checkbox at the bottom of the dialog.

When you've customized your toolbar, click **Save Preset** to save it. If you don't do that, Photoshop will still save your tools, but having the preset saved makes it easier to return to these settings in the future.

If you need to access hidden tools in the future, click the Extra Tools button at the bottom of your toolbar, as shown next.

You can click the >> symbol at the top of your toolbar to switch between a one- and two-column layout, as shown next.

Customizing Workspaces

Workspaces are custom arrangements of windows. For example, I like to see the histogram when I edit photos, so I created a custom workspace that shows the **Histogram** window. Workspaces can also include custom keyboard shortcuts, menus, and toolbars.

Creating and Saving a Workspace

To create a custom workspace, simply arrange Photoshop the way you like by displaying windows (using the **Window** menu) and dragging them into place (by dragging the tab at the top of the window).

To save your workspace so you can easily recall it, select **Window | Workspace | New Workspace**. Type a name, select all three checkboxes, and click **Save**.

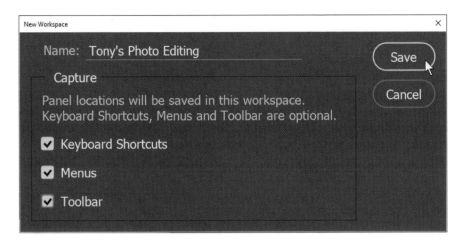

If you want to restore your workspace later (perhaps because you changed something you didn't like), you can select **Window | Workspace | Reset**.

If you make changes to your workspace that you want to make permanent (even if you choose **Window | Workspace | Reset**), select **Window | Workspace | New Workspace**. When prompted, type the exact same name as your current workspace, and Photoshop will overwrite that workspace's settings.

You can create multiple workspaces for different types of work. For example, if you use different tools for landscape and portrait photography, you could create a separate workspace for each. I prefer to use a single workspace for all types of photography, though. Changing workspaces is too disorienting for me.

Exporting and Importing Workspaces

When you create a new workspace, Photoshop saves it as a .PSW file. If you can find this file, you can transfer it to a different computer to use the same workspace. Unfortunately, Photoshop doesn't currently provide an easy way to automatically sync workspaces between different computers (for example, using Creative Cloud).

Unfortunately, it's not easy for me to tell you exactly where your workspaces files are saved. The specific folder varies depending on your username, the current version of Photoshop, and whether you're using a Mac or a PC.

The easiest way to find your workspace might be to search your entire computer for the Workspaces folder. That folder should contain one or more .PSW files. Copy those files to the Workspaces folder on your destination computer, restart Photoshop, and you should be able to use the saved Workspace.

On my Windows computer, with the current version of Photoshop, the folder is **C:\ Users\<*username*>\AppData\Roaming\Adobe\Adobe Photoshop CC 2015.5\Adobe Photoshop CC 2015.5 Settings\WorkSpaces**.

Note that transferring a workspace to another computer might not work as well as you hope. Computers might have monitors with different resolutions, which could cause the window layout to be incorrect.

20 CONFIGURING PREFERENCES

Watch training videos at:
SDP.io/PSEDL

This chapter describes Photoshop's many preferences. Most users can be happy with the defaults. However, depending on your editing style and workspace, some of these options can make your computer and editing much more efficient.

To open the Preferences dialog, select **Edit | Preferences | General** (on a PC) or **Photoshop CC | Preferences | General** (on a Mac). You can quickly open the Preferences dialog by pressing **Ctrl+K** or **Cmd+K**.

To reset your settings, hold down the **Alt/Opt** key to change the **Cancel** button to **Reset**. Then, click **Reset**.

General

Here's a quick overview of the General settings. The following screenshot shows my preferred settings; the only option I change from the default is selecting **Show "Start" Workspace When No Documents Are Open**.

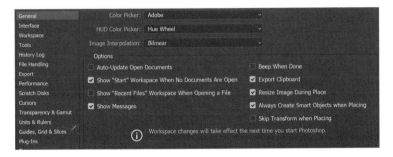

Color Picker

When you click the foreground or background color, Photoshop prompts you to pick a color using the color picker you select. The default option (shown next) is usually the best choice.

HUD Color Picker

When you're using the Brush tool, press **Shift+Alt+right-click** (on Windows) or **Ctrl+Opt+Cmd** (Mac) to display the HUD color picker. The next two examples show the Hue Strip (default, on the left) and the Hue Wheel (on the right). They're both equally functional; pick whichever style and size you prefer. You can also change the size.

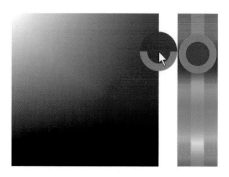

Image Interpolation

Increasing the resolution of an image doesn't increase the detail of an image. To add those extra pixels, Photoshop interpolates the extra detail, which is a fancy way of saying Photoshop makes a wild guess about what new pixels it should put between your existing pixels as it stretches your photo out.

Most of the time, when you need to increase the resolution of an image, you should use the **Image | Image Size** tool. That tool automatically selects an interpolation method. The following examples show zooming in 400% using the Image Size tool, and selecting with **Bicubic**, **Preserve Details** (which was selected automatically), **Nearest Neighbor**, and **Bilinear**:

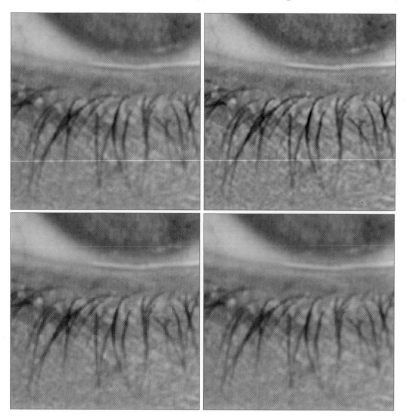

In this example, Photoshop automatically selected the best interpolation method, which means you could have gotten the best results without worrying about any of this.

Another time Photoshop interpolates is when you use the Transform tools (such as **Edit | Free Transform**). Again, Photoshop ignores the Image Interpolation preference and allows you to change the **Interpolation** directly from the **Transform** toolbar.

The Image Interpolation option doesn't change the **Image Size** or Transform tools, which are the typical way you'd scale your pictures. Even with those tools, the automatic option is usually the best. So, it doesn't make much of a difference.

Auto-Update Open Documents

If you select this checkbox, Photoshop will prompt you to reload an image that you're currently editing if you also edited and saved that file in a different app. For example, you might have been currently editing a picture.jpg in Photoshop, and then you suddenly decided to edit picture.jpg in Paint and save it. If the **Auto-Update Open Documents** checkbox were selected, Photoshop would prompt you to reload picture.jpg.

Show "Start" Workspace When No Documents Are Open

If you select this checkbox, Photoshop shows you a list of recently used files (shown next) when you open Photoshop without editing a photo. That's better than a blank screen, so I recommend selecting this option. If you change this option, you have to restart Photoshop to see the effect.

Show "Recent Files" Workspace When Opening A File

If you select this checkbox, Photoshop shows you the fairly useless Recent Files pane (shown next) when you select **File | Open**. If you change this option, you have to restart Photoshop to see the effect.

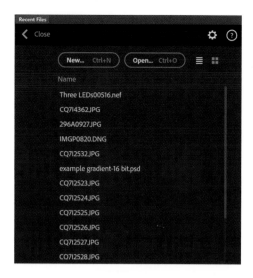

I say that this option is useless because Photoshop always lists recently used files under **File | Open Recent**. Therefore, selecting this option offers no benefit, and simply requires an extra click when opening files (because you have to also click the **Open** button).

Show Messages

Occasionally, Photoshop might show you a message about a feature that you're using. These messages are usually helpful and are easily dismissed, but if you find them annoying, you can clear this checkbox.

Beep When Done

I've never wished that Photoshop beeped more often, but even when I selected the checkbox, I couldn't find a command that would cause Photoshop to beep. You can ignore this option.

Export Clipboard

This option, selected by default, allows you to copy from Photoshop and paste into another app. This is the normal behavior for the clipboard, so you should leave it selected.

Resize Image During Place

Despite extensive testing, I couldn't determine that this setting impacted Photoshop's behavior in any way.

Always Create Smart Objects When Placing

Smart Objects are the best way to handle new layers, so it's best to leave this option selected.

If you're curious, Smart Objects use non-destructive editing. So, you can edit an image in a layer repeatedly and not have any strange artifacts appear because of the editing.

Skip Transform When Placing

When you place an image onto an existing image (for example, by dragging a second picture into an existing image), Photoshop automatically starts a free transform so you can quickly scale it to the right size. This is useful; I usually need to scale the new image, and if I don't, I can simply press Enter to skip the transform.

Therefore, I recommend leaving this option deselected. If you do select it, you can press **Ctrl+T** to perform a transform after placing an image.

Interface

The Interface section allows you to change Photoshop colors and appearance. Here's how I've configured the user interface:

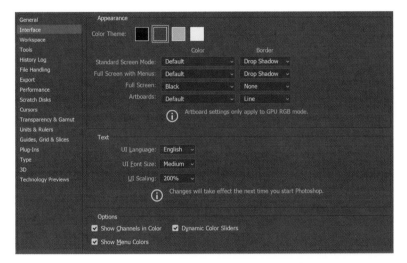

I suggest adjusting the settings in the Appearance section to make yourself happy. I use the default settings only because I have to take screenshots.

If the text in Photoshop is too big or too small, you can adjust them using the settings in the Text section. I have the **UI Scaling** set to 200% because I use a high-resolution monitor. Adjust the UI Font Size and UI Scaling so that you can read everything easily. Unfortunately, you have to restart Photoshop to see the effects of every change.

Here are three options on this page, as described in the next sections.

Show Channels in Color

If you use the Channels panel to separately view the red, green, and blue channels, Photoshop shows each channel in black and white by default, as shown next:

If you'd rather see the channels in color, select this checkbox. The channels will resemble the following screenshot:

Dynamic Color Sliders

In earlier versions of Photoshop, this option made a very subtle difference to the behavior of the color picker tool. In recent versions of Photoshop, we don't need to use the old color picker, so you can ignore this option.

Show Menu Colors

If you use **Edit | Menus** to customize the colors of your menus, clear this checkbox to hide your custom menu colors. I'm not sure why you'd go through the trouble of changing your menu colors only to hide them, but Photoshop gives you that option.

Workspace

This section describes the Photoshop Workspace preferences. The following figure shows my preferred settings.

Auto-Collapse Iconic Panels

If you select this checkbox, Photoshop automatically hide some panels when they're not in use. Specifically, if you add an adjustment layer, Photoshop will display a panel so that you can change the settings. With this option selected, Photoshop hides that panel as soon as you click elsewhere.

I prefer to have this option selected.

Auto-Show Hidden Panels

If you press Tab to hide open panels, hover your cursor over the edge of the screen where the panel is hiding to show those hidden panels. If you clear this checkbox, Photoshop will do nothing when you hover over the hidden panels. I prefer to have the option selected.

Open Documents as Tabs

When you select this option, Photoshop displays open documents as tabs:

When you have the option cleared, open documents float freely:

I suggest leaving this option selected. If you do have this option selected and you want an image to float separately in its own window, just grab the tab and drag it. Alternatively, if you clear this option, you can always drag images to the top of the screen to dock them as tabs.

Enable Floating Document Window Docking

By default, this checkbox is selected and you can drag documents to the top of the window to dock them as tabs. Clearing this option prevents you from doing this, and I can't imagine why you'd want to disable it.

Large Tabs

If you select this checkbox, Photoshop makes the document and window tabs taller. If you're having a hard time clicking tabs, this option makes it easier but wastes a bit more screen space.

The next two examples show how the tabs appear when the checkbox is cleared and selected.

Enable Narrow Options Bar

If you have a small or low resolution monitor, selecting this option will make the options toolbar less wide. It doesn't change the options toolbar for every tool, but the next two examples show the Crop options toolbar with this checkbox selected and then cleared.

As you can see from the previous example, selecting the Enable Narrow Options Bar checkbox causes Photoshop to use icons for two of the options instead of writing out the text.

It's only beneficial to enable this option if the options bar doesn't fit on your screen.

Tools

This section describes the Photoshop Tools preferences. The following figure shows my preferred settings.

Show Tool Tips

Tool tips are text messages that provide occasionally useful descriptions of user interface elements when you hover the cursor over them, as shown next. Most of the tool tips, including the following example, are pretty useless. However, some of them are helpful.

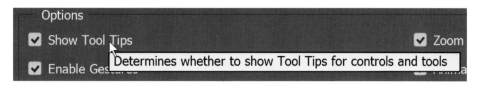

If you find tool tips annoying, clear this checkbox.

Enable Gestures

If you have a touch screen, Photoshop supports intuitive gestures, such as pulling or pinching the screen to zoom in or out. If you find those gestures annoying, clear this checkbox. It won't make any difference if you don't have a touch screen.

Use Shift Key for Tool Switch

You can select groups of tools with keyboard shortcuts. For example, if you press J, Photoshop will select the default healing tool. If you press **Shift+J**, Photoshop will select the next healing tool. So, you can press **Shift+J** multiple times to cycle through all the healing tools until Photoshop selects the tool you need.

If you don't want to press the Shift key while switching tools, clear this checkbox.

Overscroll

If you select this checkbox, you can scroll further to the side to place the visible document at the edge of your window when it's zoomed out. I can't imagine why that would ever be useful, but it doesn't really harm anything, either.

Enable Flick Panning

With this option selected, you can pan the image, and if your cursor is moving when you release the mouse button, the image will continue to scroll. This simulates inertia-like rolling dice that continue to move after you release them.

It's a minor aspect of the user interface that doesn't have much practical purpose, but for those of us accustomed to using touch interfaces on smartphones and tablets, it makes sense to leave it selected.

Use Legacy Healing Algorithm for the Healing Brush

Adobe regularly improves the healing brush algorithm. In my opinion, every update has improved the performance. If, after an update, you find that the healing brush is doing worse, you can select this checkbox.

Double Click Layer Mask Launches Select and Mask Workspace

Photoshop CC 2015.5 added the Select and Mask tool. By default, when you double-click the layer mask, Photoshop will show you that new tool. Previous versions of Photoshop showed you the fairly useless Properties dialog (shown next).

If you do clear the checkbox, you can always right-click a layer mask and then select **Select and Mask**.

Vary Round Brush Hardness based on HUD vertical movement

Photoshop offers what Adobe calls a Heads-Up Display (HUD) that you can use to adjust brush size and hardness without clicking the options bar (shown next). To use it on a PC, hold **Ctrl + Alt** while right-dragging your mouse. On a Mac, hold **Ctrl + Opt** while dragging.

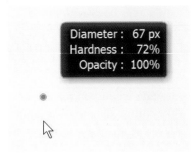

Photoshop always changes the diameter as you drag your mouse left or right. If this option is selected, Photoshop changes the hardness of the brush while you drag up or down. If you clear this option, Photoshop changes the Opacity instead of the hardness.

Snap Vector Tools and Transforms to Pixel Grid

You probably don't need to change this option. However, you should know how to configure snapping in Photoshop.

From the **View** menu, select **Snap To**, and then choose which user interface elements you want your cursor to automatically align with. If you find the snapping annoying, just select **View | Snap To | None**. The snapping only takes effect if guides or the grid are visible. To show them, select the options from the **View | Show** menu.

Show Transformation Values

When you change an image with a transform, Photoshop shows the current size to the upper-right of your cursor, as shown next. Use this option to move the display to a different corner or to hide it completely.

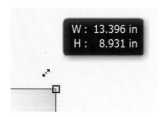

Zoom with Scroll Wheel

You can use your mouse scroll wheel to zoom in and out. If you'd rather scroll the current document up and down (or side-to-side, if your mouse wheel supports that), clear this checkbox.

Animated Zoom

When this checkbox is selected, Photoshop smoothly zooms in and out when you press **Ctrl++** or **Ctrl+-**. If you clear the checkbox, the zooms happen a bit quicker because Photoshop skips the animation.

It doesn't seem to make a difference when you zoom using the scroll wheel.

Zoom Resizes Windows

This option only makes a difference when you have an image in a free-floating window. In that case, if the checkbox is selected, Photoshop will resize the window to match the size of the image when you zoom in and out. If this checkbox is cleared (which I prefer) the window size stays the same when you zoom.

Zoom Clicked Point to Center

This option controls a subtle behavior of the user interface for backwards compatibility. When this checkbox is cleared (the default) and you use the zoom tool to zoom in, Photoshop zooms so that your cursor stays in the same location of the photo. If you select this checkbox, Photoshop instead centers the spot you clicked in the middle of the image.

If you click in the middle of your picture, the behavior will be exactly the same. If you zoom towards the edge of the picture, the picture seems to jump around when you have this option selected. I suggest leaving this option cleared.

History Log

If you select the History Log checkbox, Photoshop will save a log of your changes to the metadata of a file, to a separate text file, or both.

That could be really useful if Photoshop would read the history and allow you to undo changes you had made in a previous session. Unfortunately, it doesn't work that way; your only option is to view the changes you've made in a text file.

The Edit Log Items list gives you three choices:

- **Session**. Only saves when you open, save, and close files.

- **Concise**. Saves only the information visible in the History window.

- **Detailed**. Saves detailed information about every change, as shown next.

```
2016-06-29 17:36:13   File 296A8258.psd closed
        New Levels Layer
            Make adjustment layer
                Using: adjustment layer
                Type: levels
                Preset Kind: Default
        Modify Levels Layer
            Set current adjustment layer
                To: levels
                Preset Kind: Custom
```

```
          Adjustment: levels adjustment list
          levels adjustment
          Channel: composite channel
          Gamma: 1.74
2016-06-29 17:36:55   File 296A8258.psd closed
          Close
          257
          true
```

I don't find the History Log useful, so I don't use it.

File Handling

The File Handling tab of the Preferences dialog gives you control over how files are opened and saved. The following figure shows how I prefer to configure these settings.

File Saving Options

The next few options relate to saving files. The only option I change is to **Automatically Save Recovery Information Every 5 Minutes**.

Image Previews

Complex files, such as TIF and PSD files, can store a JPG image preview. That allows Lightroom or a file browser to generate a thumbnail or larger preview without parsing the entire file.

In other words, an image preview makes it faster to work with large image files but slightly increases the file size. I prefer to enable image previews.

File Extension or Append File Extension

On a Windows computer, you have the option to use either upper or lower case extensions. In practice, it doesn't make much difference until you upload your images to a website or a computer running a different operating system. Most people use lowercase file extensions.

On Macs, you have the **Append File Extension** option. For compatibility with Windows computers, you should leave the default **Always** option selected.

Save As to Original Folder

If you choose **File | Save As**, Photoshop will prompt you for a filename. By default, it will select the same folder the file is currently stored in, and this makes sense for most people. If you'd rather Photoshop start the browsing process from your Documents folder, clear this checkbox.

Save in Background

By default, Photoshop lets you continue working on a file while it saves it in the background. This makes a lot of sense because large files can often take a full minute to save. If for some reason you'd rather wait while Photoshop saves a file, you can clear this checkbox.

Automatically Save Recovery Information Every:

Photoshop crashes occasionally. When it does, it will attempt to recover your work, even if you haven't saved it. This doesn't overwrite your original file; that only happens when you manually save the image.

Set this to the minimum amount of work you can stand losing after a crash. I changed it to the minimum of 5 minutes.

File Compatibility

The next few options relate to file compatibility. I prefer to clear the **Ask Before Saving Layered TIFF Files** checkbox and set **Maximize PSD and PSB File Compatibility** to **Never**.

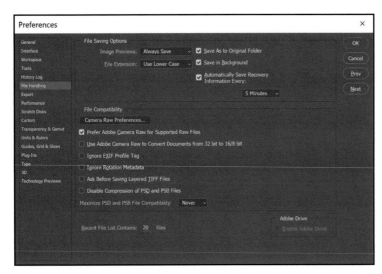

Camera Raw Preferences

Clicking this button opens the following dialog. The only non-default setting I use is to select the **Update embedded JPEG previews** checkbox and set the value to **Full Size**.

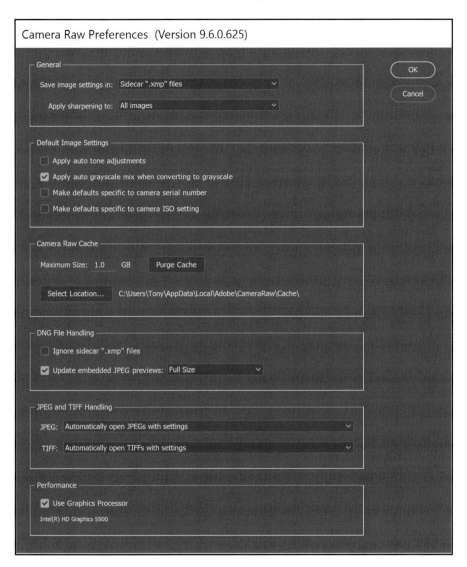

Camera Raw Preferences (Version 9.6.0.625)

The defaults work fine for most people, but here's a quick discussion of each option:

- **Save image settings in**. When you adjust the exposure or other settings of a raw file, Photoshop needs to save those settings somewhere other than the file because Photoshop won't modify the raw file. By default, it saves them as a separate .XMP file with the same name in the same folder. That's usually the better option because you can copy the raw file and the .XMP file to a different computer and keep the adjustments. If you choose to save it in the Camera Raw database, Photoshop won't generate separate files, which is nice, but your settings won't be available on other computers.

- **Apply sharpening to**. Camera Raw usually applies some sharpening to an image; this is normal and desired. If you don't like this behavior, you can have it apply sharpening to only the preview.

- **Apply auto tone adjustments**. If you select this checkbox, Photoshop will try to adjust the exposure and contrast of raw files automatically. I would never select this; sometimes the auto toning is terrible. It's easy to apply auto toning manually if you want it.

- **Apply auto grayscale mix when converting to grayscale.** If you convert a raw file to black and white, Camera Raw does something smart—it adjusts the levels of different colors so they look more like a traditional black and white film photograph. If you'd rather have the levels for all colors be flat, select this checkbox. Either way, you can adjust the colors yourself later.

- **Make defaults specific to camera serial number.** You can change the default Camera Raw processing. If you work with different cameras and want different defaults for each camera, select this checkbox.

- **Make defaults specific to camera ISO setting.** Similar to the last option, select this if you want to have different defaults for every ISO. You could, for example, increase the noise reduction for photos taken at ISO 3200. You probably don't want to select this, though, because you'll need to reset your defaults for every ISO, even ISO 100 and ISO 125.

- **Camera Raw Cache.** By default, Photoshop caches raw file data to a file to improve performance when you open the same raw file multiple times. I rarely repeatedly open the same raw file in Photoshop, so this doesn't make much difference. However, 5 GB is also not much space to dedicate to the cache, so I usually leave this enabled. If you want to save same disk space, you could reduce the cache size. If you want to clear the cache and free some disk space, click the **Purge Cache** button. Click **Select Location** to store the cache on a different drive. For best performance, choose your fastest permanently connected drive.

- **Ignore sidecar ".xmp" files**. Photoshop, by default, saves changes you make to a raw file in a .XMP file in the same folder with the same name. I don't recommend selecting this checkbox; it will result in your raw file edits disappearing the next time you open the raw file. If you want to remove the changes for a single file, you could delete the .XMP file from the folder.

- **Update embedded JPEG previews**. When you take a raw photo, your camera stores a JPG preview of that file in the image's metadata. This allows your computer to display a preview of the raw file much faster than would otherwise be possible. Photoshop can overwrite this preview with the image that it creates based on the Camera Raw default settings or your own changes. Bigger previews will be sharper. I prefer to have this checkbox selected and to use the **Full Size** preview.

- **JPEG and TIFF Handling.** Camera Raw can process JPG and TIFF files, even though they're not raw files. The default settings should be fine, but you can change them to prevent Camera Raw from processing these non-raw files.

- **Use Graphics Processor**. Most computers have a graphics processor that's really fast at doing math problems. Camera Raw uses that processor to improve performance. If Camera Raw crashes, it might be because of a flaw in your graphics processor. In this case, you could clear the checkbox. If Camera Raw seems particularly slow, you might try clearing this checkbox to see if it improves performance.

Prefer Adobe Camera Raw for Supported Raw Files

If you have some other raw processing software that you'd rather use when you open images in Photoshop, clear this checkbox. To work, the software will need to properly install itself on your computer.

Use Adobe Camera Raw to Convert Documents from 32 bit to 16/8 bit

Most image files are only 16-bit or 8-bit, and this option only impacts 32-bit files. Therefore, it probably won't make a difference for you. The default is fine for most users.

Ignore EXIF Profile Tag

EXIF profiles tell Photoshop what color space an image is. You wouldn't typically want Photoshop to ignore this. However, if you have a file with incorrect EXIF data, you might temporarily select this checkbox.

Ignore Rotation Metadata

Photoshop handles images as horizontal by default. Most modern cameras will detect if you turn the camera sideways to take a vertical shot and write that information into the image file. This allows Photoshop to automatically display vertical photos vertically.

There's really no reason to select this checkbox. Many years ago, before cameras automatically detected vertical shots, Adobe updated Photoshop to respect that information. Adobe was concerned that it might annoy some users, so they gave them the option to revert back to the older behavior.

Ask Before Saving Layered TIFF Files

If you add layers to a file, the file size can grow really large. However, it's really important to save those layers because otherwise you won't be able to edit the layers the next time you re-open the file. By default, Photoshop prompts you before saving the layers. If I didn't want to save the layers, I would manually flatten the image, so I clear this checkbox.

Disable Compression of PSD and PSB files

Photoshop uses lossless compression to reduce the file size of PSD and PSB files. It doesn't degrade image quality and it saves space, so you wouldn't want to disable it. This option only exists in case you have a problem opening the files in a non-Adobe app that doesn't support compression.

Maximize PSD and PSB File Compatibility

Every time you save a PSD or PSB file, Photoshop will prompt you to save it in an older file format that's compatible with older apps. I never have problems using the new file format, so I set this option to **Never**.

Recent File List Contains

Photoshop stores a list of recently opened files under **File | Open Recent**. This is helpful for when you want to continue working on a file. This option controls how many files Photoshop remembers. I've increased it to 20, which makes the list a little longer but still manageable.

Adobe Drive

If you work at a larger business that uses a content management system (CMS) or Adobe Experience Manager, and you have Adobe Drive installed, you can select this checkbox. Most people won't use it. For more information about Adobe Drive, visit *sdp.io/adobedrive.*

Export

This page controls the behavior for the **File | Export | Quick Export** function. For detailed information, refer to Chapter 18, "Understanding Image Files."

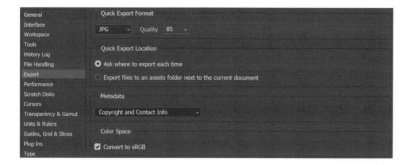

Performance

The performance settings, shown next, tweak how Photoshop uses your computer's resources.

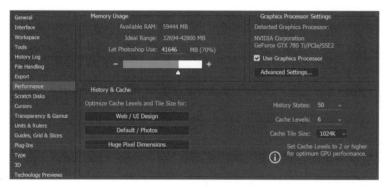

Adjust **Memory Usage** to set the maximum amount of physical memory (RAM) that Photoshop might use if it needs it. Photoshop won't always use this much memory; you're just defining an upper limit. For example, my settings allocate a maximum of 41 GB of RAM to Photoshop, but typically, it uses less than 1 GB. However, if I'm creating a large panorama, Photoshop might take advantage of my computer's extra memory.

The only disadvantage to setting the **Memory Usage** value higher is that your computer might not have more memory available to other applications when you're working with really large files in Photoshop. For best performance, set the Memory Usage value high, but close other applications while using Photoshop to edit large files or files with many layers.

Usually, Photoshop will be faster with the **Use Graphics Processor** checkbox selected. This depends on the type of graphics processor your computer has, however. You can try clearing the checkbox and seeing if Photoshop is faster.

In the **History & Cache** section, the default is probably fine. If you have a high megapixel camera, such as a 36- or 50-megapixel camera, or if you make panoramas, you might get better performance by clicking **Huge Pixel Dimensions**. Clicking any of the three buttons will adjust the **History States**, **Cache Levels**, and **Cache Tile Size** values.

Scratch Disks

Photoshop uses scratch disks to temporarily store bits of data for which it doesn't have room in memory. This data isn't permanent, and you don't need to worry about it being deleted.

For best performance, select a fast disk (such as an SSD) for your scratch disk. If you have more than one SSD, choose a different SSD than you use for your system and Photoshop.

Cursors

The Cursors section of the Photoshop preferences allows you to change the appearance of the brush cursor. I haven't found that adjusting these settings significantly changes how I work with Photoshop, so I leave them at the default.

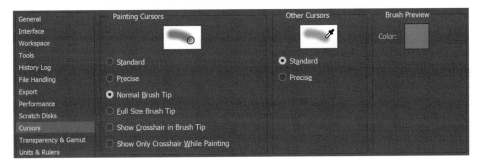

This section is unusually user-friendly for Photoshop because it shows a preview of how your settings will work. For example, if you select Show Crosshair in Brush Tip, Photoshop will demonstrate that at the top of that section (shown next). Therefore, to get the most out of this section, I suggest you open that window and try the different options.

Transparency & Gamut

If you remove part of the background layer, you'll be left with transparency. By default, Photoshop shows transparency as a grey and white checkerboard. The next figure shows a photo with the sky removed.

This checkerboard is fine. However, you can adjust the size and colors in the grid. I leave it at the default, as shown next.

Units & Rulers

The Units & Rulers preferences (shown next) don't have a huge impact on how we as photographers work with Photoshop. The **Print Resolution** and **Screen Resolution** options don't have much impact on how we actually print and view images.

However, you might want to change **Rulers** to your preferred unit of measurement, such as **Centimeters** or **Inches**. You'll only notice the difference if you turn on rulers (**View | Rulers** or **Ctrl+R**). However, that can be useful when cropping an image for a specific print size.

Guides, Grid & Slices

Another section of preferences with little practical purpose for photographers is Guides, Grid & Slices. With these tools, Photoshop will display lines over an image to help designers align columns, text, and other visual elements.

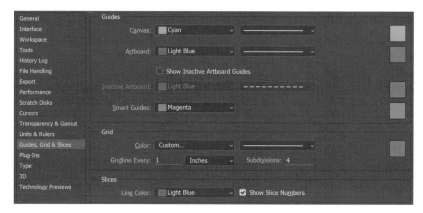

If you do need to use these tools, you might find that the colors of the lines are too similar to those in the photo, and they might be hard to see. Use this preferences section to change their color to something more visible.

You can turn on guides using **View | New Guide** (to add a single line) or **View | New Guide Layout** (to add multiple lines). Turn on the grid using **View | Show | Grid**. Slices are primarily used for Web design, so this book won't cover them.

Plug-ins

Plug-ins are extra bits of software that add functions to Photoshop. For example, you can install the free Google Nik plug-ins (*sdp.io/nik*) and use them to create special effects on your images.

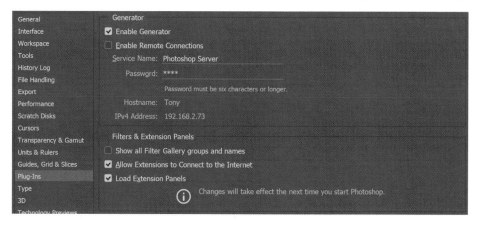

There are three options related to filters:

- **Show all Filter Gallery groups and names**. Confusingly, Photoshop gives you access to filters in two different ways: from the **Filters** menu and from **Filters | Filter Gallery**. If you want to see all the filters on the menu, even those that are also listed in the Filter Gallery, select this checkbox.

- **Allow Extensions to Connect to the Internet**. Some filters might want to connect to the Internet. For example, they might check to see if there is an update available. You can clear this checkbox to prevent Internet connections, which you might want to do for privacy or security purposes.

- **Load Extension Panels**. Extensions, including the Google Nik plug-ins, can add new panels to Photoshop. You can easily close these panels, or you can clear this checkbox to prevent them from loading.

If you're curious, the Generator tool allows designers to quickly find assets (such as stock photos) to drop them into their current project. It's not a tool most photographers will ever use.

Type

These options are of more use to designers than photographers. The only option you might want to change is **Use Smart Quotes**. By default, Photoshop will change quotes to smart quotes, which face in towards the text. "For example, these are smart quotes," Tony said, as he pointed out that the quotes before and after text are angled slightly different. If you turn this option off, the quotes will be straight quotes, and they'll be the same before and after a quote.

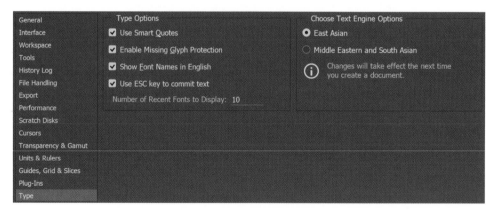

3D

Photoshop has the ability to create and edit 3D objects. For example, you could import a 3D rendering of an airplane and then insert it into a photo. Though 3D might be relevant to photographers in the future, at this moment in history, it's outside the scope of this book.